A-LEVEL YEAR 2

STUDENT GUIDE

AQA

Biology

Topics 7 and 8

Genetics, populations, evolution and ecosystems

The control of gene expression

Pauline Lowrie

PHILIP ALLAN FOR
HODDER
EDUCATION
AN HACHETTE UK COMPANY

The author would like to thank South Sefton Sixth Form College.

Philip Allan, an imprint of Hodder Education, an Hachette UK company, Blenheim Court, George Street, Banbury, Oxfordshire OX16 5BH

Orders

Bookpoint Ltd, 130 Park Road, Milton Park, Abingdon, Oxfordshire OX14 4SE

tel: 01235 827827

fax: 01235 400401

e-mail: education@bookpoint.co.uk

Lines are open 9.00 a.m.–5.00 p.m., Monday to Saturday, with a 24-hour message answering service. You can also order through the Hodder Education website: www.hoddereducation.co.uk

Contents

■ Getting the most from this book

Exam tips

Advice on key points in the text to help you learn and recall content, avoid pitfalls, and polish your exam technique in order to boost your grade.

Knowledge check

Rapid-fire questions throughout the Content Guidance section to check your understanding.

Knowledge check answers

1 Turn to the back of the book for the Knowledge check answers.

Summaries

■ Each core topic is rounded off by a bullet-list summary for quick-check reference of what you need to know.

Exam-style questions

Sample student answers

Practise the questions, then look at the student answers that follow.

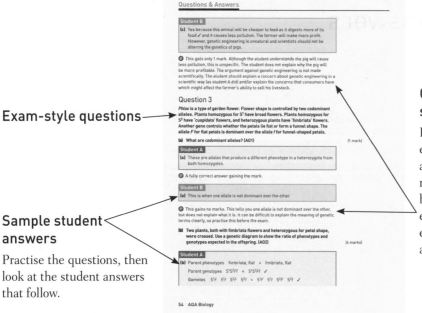

Commentary on sample student answers

Find out how many marks each answer would be awarded in the exam and then read the comments (preceded by the icon ⓮) following each student answer showing exactly how and where marks are gained or lost.

■ About this book

This guide will help you to prepare for topics 7 and 8. These topics are examined in paper 2 (together with topics 5 and 6) and in paper 3 (together with topics 1–4 and 5–6).

The **Content Guidance** covers all the facts you need to know and concepts you need to understand for topics 7 and 8. It is important that you focus on *understanding* and not just learning facts, as the examiners will be testing your ability to apply what you have learned in new contexts. This is impossible to do unless you really understand everything. The Content Guidance also includes exam tips and knowledge checks to help you prepare for your exams.

The **Questions & Answers** section shows you the sorts of questions you can expect in papers 2 and 3. It would be impossible to give examples of every kind of question in one book, but these should give you a flavour of what to expect. Two students, student A and student B, attempt each question. Their answers, and the accompanying comments, should help you to see what you need to do to score a good mark — and how you can easily *not* score a mark even though you probably understand the biology.

What can I assume about the guide?

You can assume that:

- the basic facts you need to know and understand are stated explicitly
- the major concepts you need to understand are explained clearly
- the questions at the end of the guide are similar in style to those that will appear in the final examination
- the questions assess the different assessment objectives
- the standard of the marking is broadly equivalent to that which will be applied to your answers

How should I use this guide?

The guide lends itself to a number of uses throughout your course — it is not *just* a revision aid. You could:

- use it to check that your notes cover the material required by the specification
- use it to identify your strengths and weaknesses
- use it as a reference for homework and internal tests
- use it during your revision to prepare 'bite-sized' chunks of related material, rather than being faced with a file full of notes

You could use the Questions & Answers section to:

- identify the terms used by examiners and show what they expect of you
- familiarise yourself with the style of questions you can expect
- identify the ways in which students gain, or fail to gain, marks

Develop your examination strategy

Just as reading the *Highway Code* alone will not help you to pass your driving test, this guide cannot help to make you a good examination candidate unless you develop and maintain all the skills that examiners will test in the final exams. You also need to be aware of the type of questions examiners ask and where to find them in the exams. You can then develop your own personal examination strategy. But, be warned, this is a highly personal and long-term process; you cannot do it a few days before the exam.

Things you *must* do

- Clearly, you must know some biology. If you do not, you cannot expect to get a good grade. This guide provides a succinct summary of the biology you must know.
- Be aware of the skills that examiners *must* test in the exams. These are called assessment objectives and are described in the AQA Biology specification.
- Understand the weighting of the assessment objectives that will be used. These are as follows:

Assessment objective	Brief summary	Marks in A-level paper 1/%	Marks in A-level paper 2/%	Marks in A-level paper 3/%
AO1	Knowledge and understanding	44–48	23–27	28–32
AO2	Application of knowledge and understanding	30–34	52–56	35–39
AO3	Analyse, interpret and evaluate scientific information, ideas and evidence	20–24	19–23	31–35

- Use past questions and other exercises to develop all the skills that examiners must test. Once you have developed them all, keep practising to maintain them.
- Understand where in your exams different types of questions occur. For example, the final question on A-level paper 3 will always be worth 25 marks and will test AO1 by requiring you to write an essay. If that is the skill in which you feel most comfortable, and many A-level students do, why not attempt this question first?
- Remember that mathematical skills account for about 10% of the marks. Do make sure you can carry out these calculations, including percentages, ratios, and rates of reaction.
- You need to be familiar with the techniques you have learned in the required practicals, and be able to describe how these techniques might be used in a different context. Also, you need to be able to evaluate practical investigations and data presented to you in the exam.

Content Guidance

■Genetics, populations, evolution and ecosystems

Inheritance

You will remember from the first year of your course that a gene is a length of DNA at a specific place on a chromosome, or **locus**, that codes for a polypeptide. There may be different versions of the same gene, called **alleles**, which code for slightly different polypeptides. **Diploid** organisms have two copies of each chromosome in their body cells, so they have two alleles of each gene in each cell. These alleles may be both the same (**homozygous**) or different (**heterozygous**).

The genetic make-up of an organism is called its **genotype**. The **phenotype** is the appearance of the organism resulting from the expression of its genes and its interaction with the environment.

Knowledge check 1

Match the words with their definition.

1	Locus	A	A piece of DNA that codes for a polypeptide
2	Heterozygote	B	A specific place on a chromosome where a gene is found
3	Homozygote	C	The genetic make-up of an organism
4	Gene	D	A different version of a gene
5	Genotype	E	An individual with two different alleles of a gene
6	Phenotype	F	An organism with two copies of each chromosome in their body cells
7	Allele	G	The appearance of an organism resulting from the expression of its genes and its interaction with the environment
8	Diploid	H	An individual with two identical alleles of a gene

Monohybrid crosses

Monohybrid crosses involve a single pair of alleles. In the simplest situations, one of the alleles is dominant and the other is recessive. A **dominant** allele is expressed in the phenotype even if only one copy is present. A **recessive** allele is not expressed in the phenotype unless two copies are present. We use a specific letter for one gene. We use the upper case letter to represent the dominant allele, and the lower case letter to represent the recessive allele.

It is important that you set out your working clearly in a genetics problem. For example:

In dogs, wire hair is dominant over smooth hair. A breeder crosses a heterozygous wire-haired animal with a smooth-haired animal. What proportions of hair type would you expect in the offspring?

Parent phenotype   wire haired   smooth haired

Parent genotype   Hh   hh

Gametes   H h   ×   h h

Offspring genotypes

	H	h
h	Hh	hh
h	Hh	hh

Offspring phenotypes   wire haired : smooth haired

                        2   :   2

Answer: half the offspring will have wire hair and the other half will have smooth hair.

Exam tip

There is a sequence of stages to carry out in any genetics problem.

1 Work out the parent genotypes.
2 Work out the gametes produced.
3 Set up a Punnett square with the gametes of one parent along the top and the gametes of the other parent on the side.
4 Fill in the table with the genotypes of the offspring, by pairing the gamete at the top of the row with the gamete at the side.
5 Find the ratio of genotypes in the offspring.
6 Work out the ratio of phenotypes in the offspring.
7 Answer the question you were asked.

However, some alleles do not show a straightforward dominant–recessive relationship. Alleles may be **codominant**. **Codominance** occurs where the heterozygote has a phenotype that is different from both homozygotes. We use different nomenclature for codominance. There is a capital letter to represent the gene, with the alleles in upper case letters in superscript. An example is shown below.

In shorthorn cattle, red coat colour is codominant with white coat colour. A mating between a red individual and a white individual can produce offspring that are roan, with a mixture of red and white hairs. Use a Punnett square to predict the ratio of phenotypes you would expect from a mating between a roan cow and a white bull.

We will use the symbol C^R for the red allele and C^W for the white allele. We know that the white parent is homozygous, since a heterozygote has a different phenotype from both homozygotes. The roan parent must be heterozygous. So we can set out the problem similarly to the last example.

Exam tip

If you choose your own symbols for a genetics problem, it is better to choose a letter where the upper case and lower case letters are different in shape. Here H (for hair) was chosen as H and h are different shapes. If we had used S or W, and writing the letters by hand in an exam, it would be easy to confuse an upper case letter for a lower case letter, or vice versa.

Knowledge check 2

The allele for cystic fibrosis is recessive to the normal allele. Two parents who are both normal have a baby with cystic fibrosis. What is the chance that a second baby might also have cystic fibrosis?

Parent phenotypes	roan	white
Parent genotypes	C^RC^W	C^WC^W
Gametes	C^R C^W ×	C^W C^W

Offspring genotypes

	C^R	C^W
C^W	C^RC^W	C^WC^W
C^W	C^RC^W	C^WC^W

Offspring phenotypes roan : white

1 : 1

Knowledge check 3

What are the possible phenotypes in the offspring of two roan cattle, and in what proportion would you expect them to occur?

Sex linkage

Twenty-two of the pairs of chromosomes we have in our body cells are **autosomes**, but the remaining pair are **sex chromosomes**. Males have one X and one Y chromosome, but females have two X chromosomes. The Y chromosome does not carry many genes, so sex linkage refers to genes that are present on the X chromosome. An example of a sex-linked gene is haemophilia. This is an inherited condition in which a blood clotting factor is absent, so the blood does not clot properly. The normal allele (X^H) is dominant and the allele for haemophilia is recessive (X^h). Notice the symbols used for sex linkage. They are always written as X (for the X chromosome) with a letter in superscript to indicate the allele. In this case, H for normal is upper case, as that is the dominant allele, and h for haemophilia is lower case as that is recessive. A feature of sex-linked conditions like haemophilia is that they are commoner in males than in females. This is because men have only one X chromosome, so if they have a recessive allele on the X chromosome, there is no dominant allele present to 'hide' it. Women can only have a sex-linked recessive condition like haemophilia if they inherit the allele from each parent.

Let us work through an example of sex linkage.

A woman who does not have haemophilia, but whose father has haemophilia, marries a man who also does not have haemophilia. What is the chance that they will have a son with haemophilia?

The woman does not have haemophilia, but her father did. This tells us that she must be a carrier of haemophilia as she inherited her father's X chromosome with the haemophilia allele on it. (If she had inherited his Y chromosome, she would not be female.)

Parent phenotypes	normal female, carrier	normal male
Parent genotypes	X^HX^h	X^HY
Gametes	X^H X^h ×	X^H Y

Exam tip

Examiners will tell you whether an allele is sex-linked or not. If they do not tell you it is sex-linked, then assume it is autosomal (like the vast majority of alleles). Just because a genetics question mentions a man and a woman does not mean that the gene involved is sex-linked.

Offspring genotypes

	X^H	Y
X^H	$X^H X^H$	$X^H Y$
X^h	$X^H X^h$	$X^h Y$

Offspring phenotypes   normal female : normal male : haemophiliac male

$$2 \quad : \quad 1 \quad : \quad 1$$

Therefore the chance that they will have a son with haemophilia is 1 in 4, ¼ or 25%.

(Note that if the question asked 'What is the chance that their son will have haemophilia?' it would be ½ or 50%.)

Multiple alleles

This is when there are several different alleles of a gene. Sometimes the alleles have a sequence of dominance. Codominance may be involved as well. An example is human ABO blood groups. This is determined by three alleles: I^A and I^B are codominant, with I^O being recessive to the other two alleles. There are four ABO blood groups, with genotypes as shown in Table 1.

Table 1 Genotypes of the four ABO blood groups

Blood group	Genotypes(s)
AB	$I^A I^B$
A	$I^A I^A$ or $I^A I^O$
B	$I^B I^B$ or $I^B I^O$
O	$I^O I^O$

Suppose a woman with group A has a child of group O. Her husband is group B, and cannot understand how they can have a child with a different blood group from both of them. We can explain this as follows:

The mother must have the genotype $I^A I^O$ if her child is group O, as a child of group O must receive a recessive allele from each parent. The father must have the genotype $I^B I^O$.

Parent phenotypes   group A   group B

Parent genotypes $I^A I^O$ $I^B I^O$

Gametes $I^A \quad I^O \quad \times \quad I^B \quad I^O$

Offspring genotypes

	I^B	I^O
I^A	$I^A I^B$	$I^A I^O$
I^O	$I^B I^O$	$I^O I^O$

Offspring phenotypes   one group A : one group B : one group AB : one group O

Knowledge check 4

Is it possible for a girl to inherit haemophilia? Use a Punnett square to explain your answer.

Knowledge check 5

Red–green colourblindness is a recessive sex-linked condition. A woman with red–green colourblindness has a partner with normal vision. What is the chance that they will have a child with red–green colourblindness?

Knowledge check 6

In cats, one of several genes controlling fur colour is located on the X chromosome. The gene has two versions, or alleles. One form of the gene codes for ginger fur (X^G), and the other form codes for black fur (X^B). Heterozygous female cats ($X^G X^B$) have patches of ginger and black fur, and are described as tortoiseshell. What are the possible offspring from a mating between a tortoiseshell female and a ginger male?

Knowledge check 7

Coat colour in rabbits is controlled by a series of multiple alleles as follows:

Allele	Phenotype
C	Dark grey coat
c^{ch}	Chinchilla
c^h	Himalayan
c	White

In order of dominance, they are: $C \rightarrow c^{ch} \rightarrow c^h \rightarrow c$

a Give the phenotypes produced by the following genotypes: **i** Cc, **ii** $c^{ch}c^h$, **iii** $c^h c$, **iv** cc.

b Use a Punnett square to show the possible offspring of a rabbit with genotype Cc and a rabbit of genotype $c^{ch}c$.

Dihybrid inheritance

Dihybrid inheritance involves two genes at different loci. To look at a straightforward example, let us look at a cross between a pea plant with round, yellow seeds and a pea plant with wrinkled, green seeds. Both parent plants are homozygous. Round is dominant over wrinkled, and yellow is dominant over green.

Parent phenotypes   round, yellow seeds   wrinkled, green seeds

Parent genotypes   RRGG   rrgg

Gametes   RG   RG   ×   rg   rg

Offspring genotype   all RrGg (no need to draw Punnett square)

As these are plants, we can interbreed these offspring to produce a second generation. We will now call them 'parent' plants.

Parent phenotypes   round, yellow seeds   round, yellow seeds

Parent genotypes   RrGg   RrGg

Gametes   RG   Rg   rG   rg   ×   RG   Rg   rG   rg

Offspring genotypes

	RG	Rg	rG	rg
RG	RRGG	RRGg	RrGG	RrGg
Rg	RRGg	RRgg	RrGg	Rrgg
rG	RrGG	RrGg	rrGG	rrGg
rg	RrGg	Rrgg	rrGg	rrgg

Offspring phenotypes

9 round, yellow : 3 round, green : 3 wrinkled, yellow : 1 wrinkled, green

Note that the offspring with round, green seeds or wrinkled, yellow seeds can be called **recombinants** because they have a combination of the characteristics from both the original parents. This happens because the gene for seed colour and the gene for shape of seed are on different chromosomes. They are inherited independently of each other.

Epistasis

Epistasis is when one gene locus interacts with another gene at a different gene locus. An example is flower colour in sweet pea. Two genes code for enzymes in the same pathway. The dominant allele C codes for an enzyme that produces a colourless intermediate, while the recessive allele c codes for a non-functional enzyme. Another dominant allele P codes for an enzyme that produces the pigment anthocyanin from the colourless intermediate. The recessive allele p codes for a non-functional enzyme. (Anthocyanin is a pigment that can range in colour from orange-red to violet-blue depending on the pH.)

Let us look at the outcome if we cross two heterozygous plants:

Parent phenotype   anthocyanin (coloured)   anthocyanin (coloured)

Parent genotypes   CcPp   CcPp

Gametes   CP Cp cP cp × CP Cp cP cp

Offspring genotypes

	CP	Cp	cP	cp
CP	CCPP	CCPp	CcPP	CcPp
Cp	CCPp	CCpp	CcPp	Ccpp
cP	CcPP	CcPp	ccPP	ccPp
cp	CcPp	Ccpp	ccPp	ccpp

Offspring phenotypes   9 anthocyanin (coloured) : 7 white (no pigment produced)

Note that a plant can only be coloured if it has a functional enzyme C *and* a functional enzyme P.

Knowledge check 9

In budgerigars, two genes control feather colour as shown in the table:

Genotype	Phenotype
aabb	White
aaBB, aaBb	Yellow
AAbb, Aabb	Blue
AaBb, AABb, AaBB, AABB	Green

What are the possible genotypes resulting from a cross between two heterozygous budgerigars, and in what proportion are they produced?

Linked genes

Linked genes are genes on the same chromosome. The examples above all involve genes on *different* chromosomes. When genes are on the same chromosome, they tend to be inherited together. The closer the loci of the two genes, the less likely it will be that a chiasma forms there in prophase 1 of meiosis, so crossing over is unlikely to occur to separate them.

In maize, a gene for seed colour is on the same chromosome as the gene for seed shape. Yellow (Y) is dominant over colourless (y), and smooth (S) is dominant over wrinkled (s). When a heterozygous plant (YySs) is crossed with a homozygous recessive, the offspring genotypes are as follows:

	YS	Ys	yS	ys
ys	YySs	Yyss	yySs	yyss
Proportion expected if genes were not linked	25%	25%	25%	25%
Proportion obtained	48.5%	1.5%	1.5%	48.5%

As you can see, if the genes were not linked, and each type of gamete was equally likely to occur, the outcome would be 1 yellow, smooth: 1 yellow, wrinkled: 1 colourless, smooth: 1 colourless, wrinkled. However, when this cross was carried out very few yellow, wrinkled or colourless, smooth offspring (recombinants) were obtained. This is because the heterozygous parent produced gametes that were mainly YS or ys, because these alleles were linked on the same chromosome. Very few gametes containing Ys or yS were produced, because these required a cross-over to occur between genes Y and S in prophase I of meiosis.

Exam tip

It is a good idea to revise the work you did on meiosis in year 1 of your course. You will be expected to recognise ratios that indicate linkage is occurring.

Interpreting pedigrees

Sometimes examiners will give you pedigrees to interpret. Look at the pedigree in Figure 1.

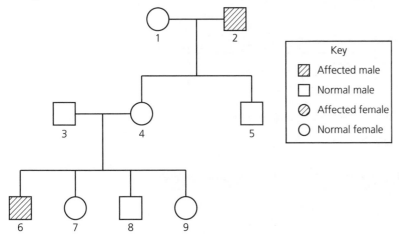

Figure 1 A human pedigree

In a pedigree like this, you may be asked 'What is the evidence that this characteristic is controlled by a recessive allele?' The word 'evidence' in a question like this really means 'proof'. You also need to refer to the individuals on the pedigree in your answer. So your answer should be '6 has the condition, but neither of his parents 3 or 4 have it' which proves the characteristic is recessive. You may also be asked to give the possible genotype(s) for individuals on the pedigree. For example, we know that 3 and 4 must be heterozygous. 7, 8 and 9 could be homozygous dominant or heterozygous.

Using the chi-squared test

The chi-squared test can be used in genetics to find out whether your results fit the ratio you would expect if your hypothesis is correct. We can work through an example.

In sweet corn, purple seed (P) is dominant over yellow (p) and smooth seeds (S) is dominant over wrinkled. Two heterozygous plants were crossed and a large number of offspring produced. See Table 2.

Table 2

Phenotype	Number
Purple, smooth	123
Purple, wrinkled	89
Yellow, smooth	71
Yellow, wrinkled	57

The student expected a 9 : 3 : 3 : 1 ratio in the offspring and wanted to see if the results obtained were consistent with this.

The null hypothesis is that the phenotypes are produced in a 9 : 3 : 3 : 1 ratio.

$$\chi^2 = \frac{\Sigma(\text{observed value} - \text{expected value})^2}{\text{expected value}}$$

The chi-squared value is best worked out using a table (Table 3).

Table 3

	O	E	(O – E)	(O – E)^2	(O – E)^2/E
Purple, smooth	123	191.25	68.25	4658.06	24.36
Purple, wrinkled	89	63.75	25.25	637.56	1.06
Yellow, smooth	71	63.75	7.25	52.56	0.82
Yellow, wrinkled	57	21.25	35.75	1278.06	60.14
Total	340				86.38

The value for chi-squared is found by adding up the last column. So $\chi^2 = 86.38$.

We look this up on a chi-squared table (Table 4).

Table 4 Chi-squared table

Degrees of freedom	Probability				
	0.9	0.5	0.1	0.05	0.01
1	0.02	0.46	2.71	3.84	6.64
2	0.21	1.39	4.61	5.99	9.21
3	0.58	2.37	6.25	7.82	11.35
4	1.06	3.36	7.78	9.49	13.28
5	1.61	4.35	9.24	11.07	15.09

Exam tip

The expected numbers are found by taking the total (340) and working out how many would be in that category if a 9 : 3 : 3 : 1 ratio was present. For example, there would be (9/16) × 340 purple smooth = 191.25.

Also notice that you can ignore any minus signs in the fourth column (O – E), since the value is going to be squared in the next column.

We need to work out the degrees of freedom. This is the number of categories minus 1. In this case we have four categories, so that makes 3 degrees of freedom. In biology, we always use a probability of $p < 0.05$. In the column for $p < 0.05$ and 3 degrees of freedom, the critical value is 7.82. Our value for χ^2 is much larger than this, so we must accept our null hypothesis. Our results do fit a 9 : 3 : 3 : 1 ratio, and the reason they are not exactly the same as the expected values is down to chance.

Summary

- The genotype is the genetic constitution of an organism.
- The phenotype is the expression of this genetic constitution. The phenotype is the result of the expression of genes and its interaction with the environment.
- An allele is an alternative form of a gene. Alleles may be dominant, recessive or codominant.
- A homozygous individual has two identical alleles of a gene, and a heterozygous individual has two different alleles of a gene at a specific locus.

- You will need to solve problems involving monohybrid and dihybrid crosses, including dominant, recessive and codominant alleles.
- Crosses may involve sex linkage, autosomal linkage, multiple alleles and epistasis.
- You should be able to use the chi-squared test to check the goodness-of-fit of experimental phenotypic ratios with expected ratios.

Populations

A **population** is all the members of one species in a particular space at a particular time that are potentially capable of interbreeding. The **gene pool** is all the alleles in a population. All the members of a population have the same genes, but they may have different alleles. Within the gene pool, each allele has a specific frequency which is a measure of the proportion of the total alleles that a specific allele contributes.

The **Hardy–Weinberg principle** is a mathematical model that predicts the frequency of alleles, genotypes and phenotypes in a population, assuming that allele frequencies do not change from generation to generation.

Diploid organisms have two copies of each gene. They will be homozygous or heterozygous for the alleles of that gene.

For example, albinos cannot make the pigment melanin, so they have pale skin, hair and eyes.

Humans who make melanin are either **homozygous** (AA) or **heterozygous** (Aa).

Albinos are homozygous for the recessive allele (aa).

Let p = the frequency of the allele A

Let q = the frequency of the allele a

Since there are only two alleles of this gene, the sum of these two frequencies must have a value of 1, i.e. the total gene pool:

i.e. $p + q = 1$

The Hardy–Weinberg principle states that:

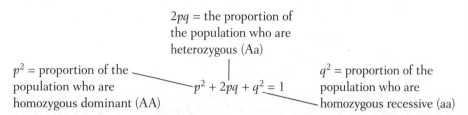

$2pq$ = the proportion of the population who are heterozygous (Aa)

p^2 = proportion of the population who are homozygous dominant (AA)

$p^2 + 2pq + q^2 = 1$

q^2 = proportion of the population who are homozygous recessive (aa)

If you know the two simple equations above, and you know what the symbols stand for, you can answer any Hardy–Weinberg problem you get.

Note that in any Hardy–Weinberg problem, you should start with the homozygous recessive individuals.

Here is an example:

In humans, tongue-rolling is determined by the dominant allele T. Non-rollers are homozygous recessive (tt). In a school, 700 students were sampled and 490 were able to roll their tongues. How many of the tongue-rollers were heterozygous for tongue-rolling? See Table 5.

Table 5 Application of the Hardy–Weinberg principle

Step	Calculation
1 Work out the frequency of homozygous recessives (this is q^2)	There were 210 (700 – 490) non-rollers, so $q^2 = 210/700 = 0.3$
2 Take the square root of the above value to get the frequency of the recessive allele (q)	$q^2 = 0.3$, so $q = \sqrt{0.3} = 0.55$
3 Find the frequency of the dominant allele (p) using the equation $p + q = 1$	$p + 0.55 = 1$, so $p = 1 - 0.55 = 0.45$
4 Put these values of p and q into the Hardy–Weinberg equation to get the genotype frequencies.	Frequency of tt = $q^2 = 0.3$ Frequency of TT = $p^2 = 0.452 = 0.20$ Frequency of Tt = $2pq = 2 \times 0.45 \times 0.55 = 0.50$ (This is what you were asked to find.)
5 Check that the frequencies total 1, to make sure you are right.	$0.3 + 0.2 + 0.5 = 1$, so we know that our calculation is correct.

There are some assumptions that we make when we apply the Hardy–Weinberg principle:

- Organisms are diploid.
- Only sexual reproduction occurs.
- Mating is random.
- Population size is large.
- Allele frequencies are equal in the sexes.
- There is no migration, mutation or selection.

Knowledge check 10

In humans, unattached earlobes are dominant, and attached earlobes are recessive. In China, it is reported that 64% of the population exhibit unattached earlobes. What percentage of the population are heterozygous?

Summary

- A population consists of all the organisms of one species in the same place at the same time, that are potentially able to interbreed.
- The gene pool consists of all the alleles in a population.
- The Hardy–Weinberg principle can be used to predict the phenotypes and genotypes in a population, and the allele frequencies.

Evolution may lead to speciation

The individual organisms within a population may show considerable variation in their phenotypes. This variation results from genetic and environmental factors. Mutations (which you learned about in year 1 of your course) are the main source of genetic variation. Variation is also the result of meiosis and the random fertilisation of gametes during sexual reproduction.

Variation is the raw material for natural selection. The members of a population compete with each other for resources such as mates, breeding sites and food. Those individuals with phenotypes that make them best adapted to the conditions are more likely to survive, reproduce and pass on their alleles. Similarly, the members of a population may be subjected to predation, or disease. Those individuals with phenotypes that give them a selective advantage, e.g. resistance to disease, are more likely to survive and pass on their favourable alleles to their offspring. As a result, the allele frequencies will be different in the next generation, with favourable alleles increasing in frequency.

This natural selection may be one of three kinds:

Directional selection is shown in Figure 2.

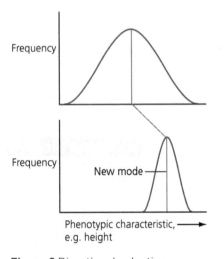

Figure 2 Directional selection

This kind of selection usually occurs when there is a change in the environment. This means that a phenotype at one extreme of the range of phenotypes is selected for, and the other extreme is selected against. As a result, the mean value for that phenotype

Knowledge check 11

Describe two ways in which meiosis contributes to variation.

Exam tip

Remember these points as the nonsense word 'VMSRAF':

There is **V**ariation in the original population. A **M**utation may occur that confers an advantage on an individual. This individual is more likely to **S**urvive, **R**eproduce and pass on the favourable **A**lleles to the next generation. This means that the favourable allele increases in **F**requency.

in the population shifts. For example, on a rocky shore where there are many crabs, dog whelks with thicker shells will be more likely to survive as their shells will be harder for the crabs to break. As a result, thicker-shelled dog whelks are more likely to survive and pass on their alleles to their offspring. Over time, the alleles for a thicker shell will increase in frequency in the population.

Stabilising selection is more likely to occur when the environment is not changing, so the population is well adapted to the environment. You can see this in Figure 3.

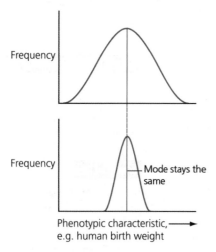

Figure 3 Stabilising selection

An example of stabilising selection is the agouti fur colour of wild rabbits. This fur colour is excellent camouflage and makes it harder for predators to spot the rabbits. Rabbits with significantly darker or lighter fur colour would be more easily seen by a predator, and therefore they would be less likely to survive to pass on their alleles for fur colour to their offspring.

Disruptive selection is shown in Figure 4.

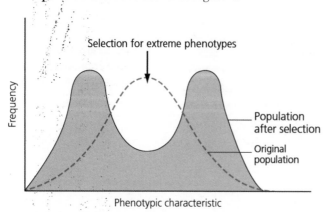

Figure 4 Disruptive selection

In this kind of selection, both extremes are selected for. For example, both pale-coloured and dark-coloured snails might be camouflaged in a particular environment, but not intermediate-coloured snails.

Knowledge check 12

a The evolution of antibiotic resistant bacteria is an example of directional selection. Explain how.

b Human birth weight is an example of stabilising selection. Explain why.

Evolution occurs as a result of the changing allele frequencies in a population. A population may become divided, for example by a geographical barrier such as a river. The gene pools in the two areas may be different. Over time, further changes occur in the gene pools as a result of natural selection in each environment. For example, the vegetation in each area may be different, so the camouflage that is suitable for an animal population in one area may be different from the most successful camouflage in the other area. Over a very long time, so many changes occur in one population compared to the other that they become reproductively isolated. For example, the population in one area may have a courtship ritual that is slightly different from that in the other area. At this point, even if the geographical barrier is removed, the two groups can no longer interbreed and produce fertile offspring. They are now regarded as separate species.

Sympatric and allopatric speciation

There are two types of speciation: sympatric and allopatric.

Sympatric speciation

- A population of organisms, e.g. a plant species, is growing and reproducing in a particular area.
- By chance, one parent produced polyploid offspring. This means that these offspring have an additional set of chromosomes.
- The polyploid offspring cannot interbreed with the other organisms, so there are now two groups of organisms with separate gene pools.
- As the two groups can no longer interbreed to produce fertile offspring, they are now regarded as separate species.

Allopatric speciation

- A population of organisms is growing and reproducing in an area.
- The population becomes divided into two smaller populations by a geographical barrier, e.g. a river or mountain range.
- The gene pools of the two populations are slightly different, by chance.
- Mutations occur in each population, and natural selection favours different phenotypes in each population.
- Eventually a mutation occurs in one population that affects reproduction, for example, the time of year at which they are fertile or some aspect of their courtship behaviour.
- Now that the two populations are reproductively isolated, they are regarded as separate species, as they cannot reproduce to produce fertile offspring even if the two populations join each other again.

Genetic drift

In small populations, there is a random element called genetic drift that affects allele frequencies. One example is a **genetic bottleneck**, which is when most of a population dies, either through disease, climatic changes, predation or hunting. The organisms that survive this near extinction will, by chance, have a much smaller variety of alleles than the original population, and the allele frequencies will be different. Even if the population increases again, the variety of alleles will be much smaller than in the original population, and the frequencies of these alleles may be very different. Another example is the **founder effect**. This is when a small proportion of a population colonise a new area. These few organisms again have a much smaller variety of alleles than the original larger population. Even when there is a much larger population in the future, they will still have a very different range of alleles, and frequency of those alleles, than the original large population had.

Knowledge check 14

The Afrikaner population of Dutch settlers in South Africa is descended mainly from a few original colonists. The present-day Afrikaner population has an unusually high frequency of the allele that causes Huntington's disease. Suggest how this has happened.

Summary

- Natural selection means that those organisms with phenotypes that are advantageous in an environment are more likely to survive, reproduce, and pass on their alleles to their offspring.
- Organisms within a population compete with each other for survival, and are subject to disease and predation.
- These are all factors that can result in differential survival and reproduction.
- As a result of differential survival and reproduction, allele frequencies will change within a population.
- Natural selection may be stabilising, disruptive or directional.

- Evolution is a change in the allele frequencies within a gene pool.
- If a population becomes separated into two different populations, over a long period of time differences in the gene pools of these populations will accumulate.
- Eventually, if these genetic differences lead to reproductive isolation, these two populations will become different species.
- Speciation may be sympatric or allopatric.
- In small populations, genetic drift may be an important factor in bringing about changes in allele frequencies.

Populations in ecosystems

A **community** consists of all the populations in a given area at the same time. The community, together with all the non-living components of its environment, make up the **ecosystem**. Ecosystems may be very large — such as a forest, ocean or tundra — but it might be very small, such as a compost heap in a garden or a rock pool. A **habitat** is the place where an organism lives, e.g. a pond. Within its habitat, every species occupies a **niche**. An organism's niche is the place where it is found and what it does there. In other words, it is its role in the ecosystem. The niche includes all the organism's interactions with its biotic and abiotic environment. For example, harvest mice live in long grass in fields, roadside verges and hedgerows where they eat seeds, berries and insects along with some roots, moss and fungi. They are mainly found in the south of England and midlands where it is warmer and drier. They are prey for many other animals including weasels, stoats, foxes, owls, hawks, crows and cats.

Populations within an ecosystem can only reach a certain size, called its **carrying capacity**. This is because the number of individuals in a population becomes limited by abiotic factors or biotic factors such as disease, predation or competition. If a population rises above the carrying capacity, the numbers will be reduced until the population is at or below its carrying capacity. Therefore the number of organisms in a population fluctuates about its carrying capacity.

Biotic factors affecting population size

Intraspecific competition is competition among members of the same species, for resources such as food supply or nesting sites (for animals) or light intensity or soil nutrients (for plants). **Interspecific competition** is competition between members of different species.

Predation is another factor that affects population size. Look at Figure 5.

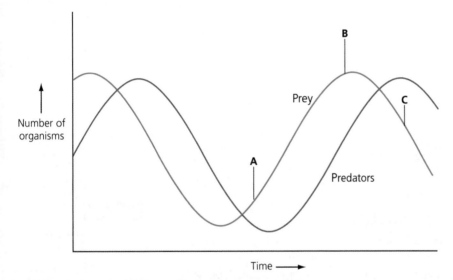

Figure 5 The relationship between a predator and its prey

- At point A on the graph, the prey population is increasing. This is because the predator population is low, so fewer prey are being eaten. Also the prey population is low, so there is little intraspecific competition for resources such as food supply.
- At point B on the graph, the prey population is high, so there is a great deal of intraspecific competition. Also, the predator population is higher, as there are many prey for them to eat. Therefore the prey population starts to decline.
- At point C, the prey population is declining because there is a large predator population, so more prey are being eaten. However, once the prey population falls to a certain level, there is intraspecific competition within the predator population for food, and the predator population gradually declines.

This cycle continues, with the predator and prey population controlling the size of each other's population.

Knowledge check 15

Paramecium is a single-celled protoctist that lives in freshwater and feeds on bacteria. The graphs below show the population growth of two species of *Paramecium* when grown separately and when grown together. Explain these results.

Species cultured separately

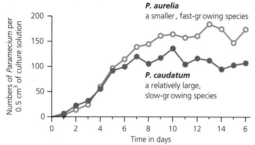

Species cultured together

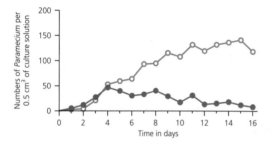

Estimating the size of a population

The size of a population of plants and other organisms that are sessile, i.e. do not move around, can be estimated using quadrats. A **frame quadrat** is a frame, usually $0.5\,m^2$, that is placed on the ground and the organisms within it counted. A **point quadrat** is a rod, usually 0.5 m long with pins every 10 cm along it. The pins are pushed down and the organisms that the pins touch are recorded. You can see these in Figure 6.

As you can see in Figure 6, these quadrats may be used to record population density, the frequency of different species, or percentage cover.

If you want to estimate the size of a population in a large, relatively uniform area, it is not necessary to count all the individuals. One way to do this is to place quadrats at **random** using coordinates generated using a random number table. You can see how this is done in Figure 7.

Alternatively, the area may be divided into a 'grid' as in Figure 6, but the student samples, say, every fifth or sixth square on the grid. In this case, sampling is **systematic** and not random, but this is still a suitable method to use.

If the area that is sampled is not fairly uniform, and the environment changes, a **transect** is used. This is particularly useful in situations such as a seashore, or from the edge of a pond towards its surroundings. A line is placed across the area using string or a measuring tape. In a line transect, all the organisms touching the line are

Exam tip

Placing quadrats at random means that each part of the area has an equal chance of being selected. It avoids bias.

Knowledge check 16

Why is closing your eyes and throwing a quadrat over your shoulder *not* a suitable sampling method to use?

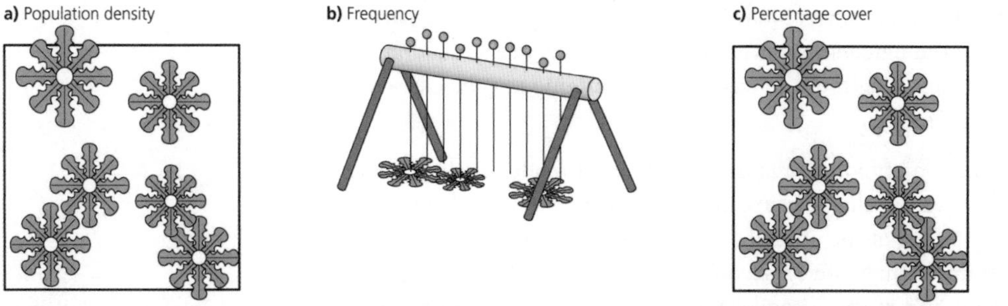

This quadrat measures 0.5 m × 0.5 m. It contains six dandelion plants. The **population density** of dandelions would be 24 plants per m^2. To get a reliable figure you would need to collect the results from a large number of quadrats. If a plant lies partly in and partly out of the quadrat, we normally count it if it overlaps the north or west side of the quadrat, and don't count it if it overlaps the south or east side.

This point quadrat frame is being used to measure **frequency**. The pins of the frame are lowered. Suppose three out of ten pins hit a dandelion plant. The frequency of dandelion plants will be three out of ten, or 30%.

Percentage cover measures the proportion of the ground in a quadrat occupied by a particular species. The percentage cover of the dandelions in this quadrat is approximately 40%.

Figure 6 Using a frame quadrat and a point quadrat

1 A map of the habitat (e.g. meadowland) is marked out with gridlines along two edges of the area to be analysed.

2 Coordinates for placing quadrats are obtained as sequences of random numbers, using computer software, or a calculator, or published tables.

3 Within each quadrat, the individual species are identified, and then the density, frequency, cover or abundance of each species is estimated.

4 Density, frequency, cover, or abundance estimates are then quantified by measuring the total area of the habitat (the area occupied by the population) in square metres. The mean density, frequency, cover or abundance can be calculated, using the equation:

$$\text{population size} = \frac{\text{mean density (etc.) per quadrat} \times \text{total area}}{\text{area of each quadrat}}$$

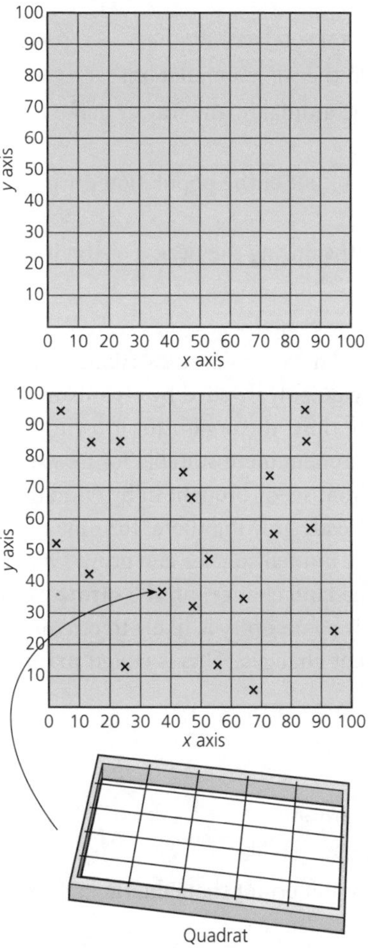

Figure 7 Placing quadrats using random number coordinates

recorded. Alternatively, a belt transect may be carried out by placing a quadrat against the line and recording the percentage cover of all the species in the quadrat. This may be done for the whole length of the transect, or sampling may be carried out at intervals of, say, every 5 metres.

Estimating the population of motile organisms, such as animals, is a little more complicated. For this, we use the **capture-mark-recapture** method. This technique involves capturing a sample of the animals, marking them harmlessly, and then releasing them. A suitable time later, a second sample of animals is captured. The number of marked animals in the second sample and the total number in the second sample are noted. The total size of the population can be calculated, assuming that the proportion of marked animals in the second sample is the same as the proportion of marked animals in the total population.

$$\frac{\text{number of marked animals in second sample}}{\text{total number of animals in second sample}} = \frac{\text{number of marked animals in population}}{\text{total number of animals in population}}$$

Rearranging this gives:

$$\text{total population} = \frac{\text{number of animals}}{\text{marked and released}} \times \frac{\text{total number in second sample}}{\text{number marked animals in second sample}}$$

However, this equation makes several assumptions:

- The animals all come from the same population.
- Marking does not harm the animal in any way or make it more likely to be seen by a predator.
- There is no migration into or out of the population during the period of the investigation.
- There are no births or deaths during the period of the investigation.

Succession

When new land is exposed — for example, a new island formed as a result of a volcanic eruption — it is very quickly invaded by organisms that colonise the land and start to grow. Over a period of time, the organisms growing on the land bring about changes that make the environment more suitable for new organisms to grow. More plants may be able to grow, from seeds brought in by wind or dropped by birds. These outcompete the organisms already growing there, forming new communities. As more plant species grow, more animal species can grow. Over a long period of time, the communities found on the land change, until a **climax community** is reached, usually mixed woodland. This community is likely to remain stable for a long period of time unless the environment changes. This is called **primary succession**. This is shown in Figure 8.

During the period of succession:

- Species diversity increases.
- A wider range of niches is present.
- The community becomes more stable.
- Biotic factors become more important than abiotic factors in affecting the survival of organisms.

Knowledge check 17

A biologist wanted to find out how many carp were in a pond. In the first sample 23 fish were caught and small tags attached. In the second sample 28 fish were caught, of which 9 had tags. How many carp are in the pond?

Knowledge check 18

A student sampled equal volumes of soil from two different fields and counted the number of worms in each. The results were:

Field A — 143 worms

Field B — 207 worms

a Give a suitable null hypothesis.

b Use the chi-squared test (see p. 14) to find out whether there is a significant difference in the population size of earthworms between these two fields.

A **xerosere** = succession under dry, exposed conditions where water supply is an abiotic factor limiting growth of plants, at least initially.

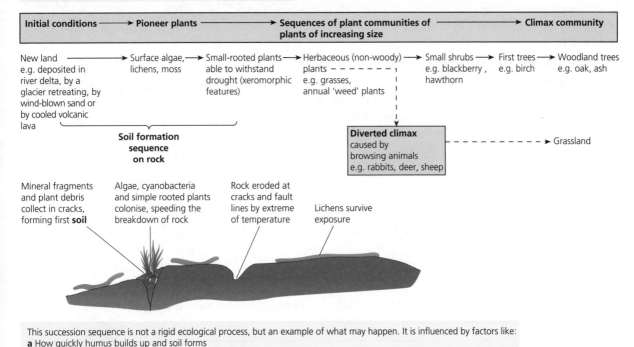

This succession sequence is not a rigid ecological process, but an example of what may happen. It is influenced by factors like:
a How quickly humus builds up and soil forms
b Rainfall or drought, and the natural drainage that occurs
c Invasions of the habitat by animals and seeds of plants

Figure 8 A primary succession on dry land

However, sometimes a climax community does not develop. This may be caused by a factor such as grazing animals. For example, the South Downs remain as grassland because of constant grazing by sheep. Sometimes habitats are actively managed to prevent succession occurring. Some grassland habitats are grazed, or cut at specific times of the year, to maintain particular species of wildlife and also insects such as butterflies. Reed beds may be cut to prevent scrub developing, and allowing species adapted to these habitats to survive. Heather moorland is often managed by controlled burning to conserve the habitat of game birds such as the red grouse.

Secondary succession occurs when an established community is suddenly destroyed, e.g. by a forest fire. This will go through the stages described earlier, but usually more quickly, since soil will already be present.

Exam tip

The examiners may give you data or information that requires you to make a judgement about the conflict between human needs and conservation. You may also be asked about the conservation of species and habitats where there is conflicting evidence. This means you will need to understand the principles in this revision guide well enough to apply them to a new situation.

Required practical 12: Investigating the effect of environmental factors on the distribution of species

This practical requires you to investigate the effect of a named environmental factor on the distribution of a given species.

Investigating the effect of soil pH on the height of buttercups

A student measured the height of 10 plants growing in different parts of the same field. She took a sample of soil adjacent to each plant and measured the pH. Her data are shown in Table 6.

She carried out a suitable statistical test on these results.

1 Suggest a suitable method the student could have used to sample the buttercups in the field.

2 Suggest a suitable null hypothesis for this investigation.

3 Suggest a suitable statistical test and calculate the test statistic.

4 The critical value for $p = 0.05$ was 0.74. Use this information to evaluate the test statistic you have calculated.

5 Give two variables, other than pH, that might affect the height of buttercups in this field. How could the student improve this investigation to take account of them?

Table 6

Height of plant/cm	Soil pH
19.0	7.1
16.5	8.0
18.0	7.2
21.0	7.4
29.5	6.8
27.5	6.3
11.0	7.6
22.5	6.4
13.5	6.8
19.0	6.0

Summary

- A population consists of all the organisms of one species in a given area at a given time.
- A community consists of all the populations in an area.
- An ecosystem consists of the community and the non-living components of the environment.
- Within its habitat, every organism occupies a niche to which it is adapted.
- An ecosystem will support a certain population size of each species, called the carrying capacity.
- The carrying capacity is controlled by abiotic factors, and biotic factors such as competition and predation.
- Quadrats may be used to sample populations by placing them randomly or systematically in an area.
- If there is a change in the environment being sampled, a transect line may be used.
- Motile organisms may be sampled using the capture-mark-recapture method.
- Primary succession occurs when pioneer species colonise an area of new land.
- These bring about changes to the environment, making it more suitable for other species to grow there.
- Over time, the environment becomes less harsh, and newer species outcompete the earlier community.
- A series of different communities develop over time, culminating in a stable climax community.
- Sometimes habitats are managed for conservation, so that succession is not allowed to progress to a climax.

The control of gene expression

Alteration of the sequence of bases in DNA can alter the structure of proteins

A mutation is a change in the base sequence of DNA. This can occur during DNA replication. Gene mutations include addition, substitution, inversion, duplication, and translocation of bases. These are shown in Figure 9.

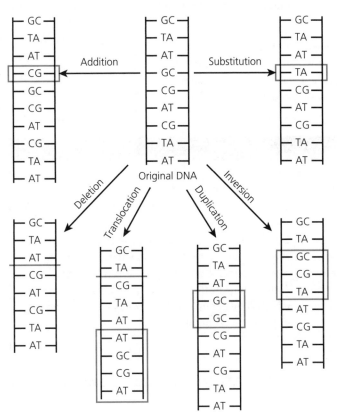

Figure 9 Different kinds of gene mutation

As you can see from Figure 9:

- **Addition** is when one or more base-pairs are added. This usually changes the amino acids coded for from the point of mutation. The triplets are changed from this point on, (or downstream from the mutation), so we call this a **frame-shift** mutation. This usually leads to a non-functional protein being coded for.
- **Substitution** is when a base-pair is changed for another base-pair. This usually changes just one triplet, so only one amino acid in the protein is changed. Nevertheless, this can still mean that a much-altered or non-functional protein is coded for. However, if the substitution affects the third base in a triplet, this may not change the protein coded for at all, since the genetic code is **degenerate**.
- **Deletion** is when a base-pair is missed out. This also causes a frame-shift mutation.

Knowledge check 19

In what part of the cell cycle does DNA replication occur?

- **Inversion** occurs when one or a few base-pairs break off, rotate, and re-insert into the DNA. This will change at least one codon and, unless the inversion affects three base-pairs, it will cause a frame-shift. Therefore the protein coded for is likely to be highly altered or non-functional.
- **Duplication** occurs when a base-pair is copied. This causes a frame-shift mutation so, once more, this is likely to produce a highly altered or non-functional protein.
- **Translocation** is when a sequence of base-pairs moves to a different part of the same DNA molecule. This will change the order of amino acids in the resulting protein significantly, and once again this is likely to cause a non-functional protein to be produced.

DNA replication is extremely accurate, but despite this, gene mutations like these occur randomly and spontaneously. However, the rate at which mutations occur can be increased by **mutagens**. Mutagens include certain chemicals, such as benzene or formaldehyde, and certain kinds of radiation, e.g. ultra-violet and X-rays.

> **Knowledge check 20**
>
> Explain why a substitution mutation may not cause any change in the protein coded for.

> **Knowledge check 21**
>
> Even if just one amino acid in a protein is changed, the protein may be non-functional. Use your knowledge of protein structure to explain why.

Summary

- Gene mutations sometimes occur during DNA replication.
- A mutation is a change in the base sequence of DNA.
- There are different kinds of gene mutation, including addition, deletion, substitution, inversion, duplication and translocation of bases.
- Most mutations result in a non-functional protein because the mutation changes at least one amino acid in the polypeptide chain.
- Some mutations only change one amino acid in the polypeptide chain, but even this change can result in a much-changed or non-functional protein.

- However, some substitution mutations do not change the protein coded for at all, because it changes the triplet to a different triplet that codes for the same amino acid.
- Some mutations change all the triplets from the point of the mutation onwards. This kind of mutation is called a frame-shift mutation.
- Although mutations occur randomly, the rate at which they occur can be increased by mutagens, such as benzene or X-rays.

Gene expression is controlled by a number of features

Most of a cell's DNA is not translated

Totipotent cells are cells that can mature into any kind of specialised body cell or placenta cells. Totipotent cells can divide to form a whole new organism. They are found in very early mammalian embryos (in the first couple of cell divisions after fertilisation). However, after this stage some of the genes in a cell become 'switched off', so they are not translated into RNA and proteins any more. When this happens, a cell becomes specialised. Some of the cells in the embryo become placenta cells and others start to develop into the fetus. After this very early stage, cells are **pluripotent**. Pluripotent cells can become any kind of specialised body cell, but they cannot divide to form a whole organism. Scientists have also learned how to convert stem cells found in adult tissue into pluripotent stem cells. These are then called **induced pluripotent stem cells**.

Adult tissues also contain **multipotent stem cells**. These cells can divide into many different specialised cell types, but not all. For example, there are multipotent cells in the brain that can divide to form neurones or **glial** cells (cells found in brain tissue), but they cannot divide to form muscle cells or blood cells. Similarly, there are multipotent cells in the bone marrow that can divide to form the different kinds of blood cells, but cannot form other cell types, such as neurones or muscle. **Unipotent** cells are stem cells that can divide to form just one kind of cell. For example, there are cells in the epithelium of the skin that can divide to produce new skin cells. Another example is cardiomyocetes, which can divide to form new heart muscle cells.

Unipotent stem cells can be changed into induced pluripotent cells (iPS) by adding appropriate protein transcription factors. This means that genes which had been 'switched off' can be 'switched on' again. The different kinds of stem cell are shown in Figure 10.

Knowledge check 22

Bone marrow transplants are sometimes given to help people produce functional blood cells. Suggest what kind of cells these transplants contain — totipotent, pluripotent, multipotent or unipotent? Explain your answer.

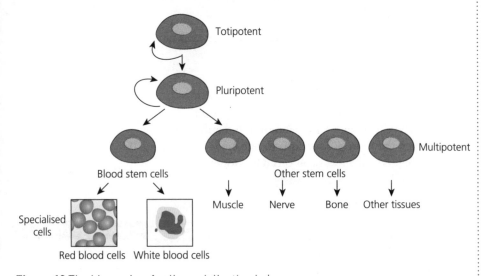

Figure 10 The hierarchy of cell specialisation in humans

Stem cells can be used to treat human disorders. Even unipotent cells have uses: for example, by taking samples of skin cells from a person's undamaged skin, sheets of skin can be grown to treat a burn on another area of the body. As these skin cells will be from the person's own body, there are no problems with rejection by the immune system. However, pluripotent stem cells are the most useful in treating disorders. Using stem cells to treat disorders is still in its developmental stages. However, scientists have injected stem cells into the retina of the eye, where they have developed into receptor cells. There is hope that stem cells will soon be useful in treating the damage caused to heart muscle by myocardial infarctions, in treating Parkinson's disease, and in producing new healthy pancreas cells that secrete insulin, to treat diabetes.

There are issues involved with using stem cells from embryos. Some people believe that destroying an embryo is equivalent to an abortion and the destruction of a potential human life. This is particularly true of some religious groups. Other people say that many embryos do not develop through to the end of pregnancy naturally, and point out that the embryos that are used are those produced for IVF (in-vitro fertilisation) treatment, and many of these were never going to develop anyway. Another problem is that stem cells from embryos can trigger an immune response in the recipient and therefore rejection. These problems are largely overcome if a person's own stem cells are induced to become pluripotent.

Summary

- Totipotent cells can mature into any type of specialised cell. These are found in very early embryos.
- Pluripotent cells can become any type of specialised body cell.
- Multipotent cells can divide into a limited range of cell types, while unipotent cells can only become one kind of specialised cell.
- When cells become specialised, some of the genes they contain are 'switched off' so that they are no longer transcribed.

- Pluripotent cells are useful in treating human disorders.
- There are some issues involved with obtaining these cells from embryos.
- However, scientists are now able to convert unipotent cells into induced pluripotent stem (iPS) cells by using protein transcription factors. iPS cells overcome many of the issues involved when using embryonic stem cells.

Regulation of transcription and translation

As was mentioned in the previous section, only a small proportion of the genes are active in a specialised cell. Different genes are active in different kinds of specialised cell, and also at different times in the same cell. Genes have **promoter** and **enhancer** regions, close to the coding DNA to which the RNA polymerase enzyme binds when it starts transcribing a gene. These promoter and enhancer regions can be switched 'on' or 'off'. If this did not happen, genes would be active or inactive all the time.

In eukaryotic cells, the transcription of specific genes may be stimulated by specific **transcriptional factors** that move from the cytoplasm to the nucleus. These protein transcriptional factors are present in the cytoplasm. They bind with the promoter and enhancer regions of the DNA. RNA polymerase recognises the 'promoter/transcriptional factor complex' and transcribes the gene. You can see this in Figure 11.

Knowledge check 23

The promoter and enhancer regions are described as being upstream of the gene they regulate. Use Figure 11 to explain why.

Knowledge check 24

Use your understanding of biological principles to suggest why transcription factors enable RNA polymerase to bind to a promoter region.

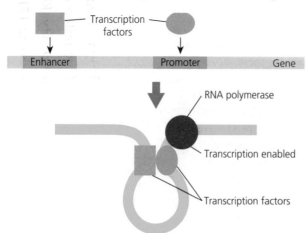

Figure 11 Transcription factors allow RNA polymerase to bind to, and transcribe, a gene

The hormone oestrogen binds with receptors in certain cells to form an oestrogen–receptor complex. This complex acts as a transcriptional factor that binds to the promoter regions of genes that stimulate cell division. This allows the hormone to have its effects, such as division of cells lining the uterus and division of cells in breast tissue. You can see this in Figure 12.

Knowledge check 25

Explain why oestrogen only stimulates cell division in certain types of cell, such as those lining the uterus, but not others.

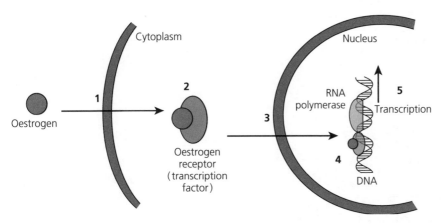

Figure 12 How oestrogen stimulates transcription

Oestrogen is lipid soluble, so it diffuses through the cell membrane (1). It binds to a protein transcriptional factor in the cell cytoplasm (2). The oestrogen-transcriptional factor passes through the nuclear envelope (3) and binds to the promoter region of specific genes (4). This leads to transcription of the gene (5).

Genes can also be 'switched off' when a **repressor** molecule binds to the promoter region and prevents transcriptional factors from binding. When the repressor molecule is bound to the promoter region, RNA polymerase cannot bind and transcribe the gene. The reason why different cells express different genes is because they contain different transcription factors and different repressors.

Genes can also be **silenced**. This is different from repressing the action of the gene because in this case the gene is still active. Cells can 'silence' a gene by breaking down the mRNA that has been transcribed from it. Therefore, although the gene is still active, no protein is made because no mRNA reaches the ribosomes for translation to take place. Gene silencing by degrading mRNA is one example of **post-transcriptional repression**.

One method of gene silencing is shown in Figure 13. This involves the action of **short interfering RNA (siRNA)**.

- Double-stranded RNA (dsRNA) is produced in the nucleus from certain genes.
- It is cut into short sequences — the siRNA — by an enzyme called 'dicer.' Short interfering RNA is very short — only 21–23 nucleotides long — and double-stranded.
- The siRNA strands separate. The antisense strand of the siRNA binds to the mRNA it is to silence.
- A complex of molecules, called RISC, binds to the siRNA/mRNA combination.
- The siRNA and RISC complex cuts the mRNA into smaller sections so that the mRNA cannot be translated.

Knowledge check 26

Is a silenced gene
a being transcribed,
b being translated?

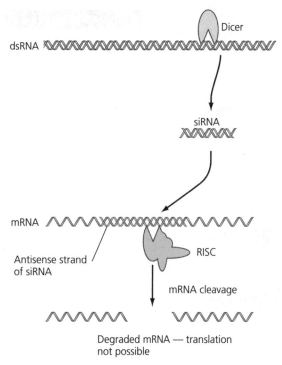

Figure 13 Gene silencing by siRNA

Epigenetic changes are heritable changes in gene function that do not involve changes in the base sequence of DNA. These changes are caused by environmental changes.

■ DNA may become **methylated**, which involves the addition of methyl groups to bases, usually cytosine bases. These chemical 'tags' are added throughout life in response to environmental factors, such as diet, stress, smoking or exercise. They may also result from signals from neighbouring cells or even the same cell. This methylation is an important mechanism for switching genes 'on' or 'off' during embryonic development, when cells are differentiating.

■ Histone proteins may become **acetylated**. You may remember that DNA is wound around histone proteins to form chromosomes. When acetyl groups are added to these histone proteins, the DNA is wound less tightly round the proteins, making it easier for RNA polymerase and transcriptional factors to bind to the DNA. On the other hand, removing acetyl groups from the histone proteins makes the DNA wind more tightly round the histone proteins, so transcription is suppressed.

Cancer occurs when cell division (mitosis) is uncontrolled. Proto-oncogenes are genes that regulate cell division while tumour suppressor genes slow down the rate of cell division. Proto-oncogenes may undergo epigenetic changes and become inactivated by methylation. This can lead to a faster rate of cell division. Other epigenetic changes may be involved in cancer development. For example, epigenetic changes in certain genes can increase the chance of cancer metastasising (spreading to other parts of the body). Genes that code for the enzymes involved in acetylation of histone proteins may be modified by epigenetic changes. This can lead to the development of cancer. Scientists are learning more about epigenetics all the time, and how this can affect cancer. Drugs that inhibit DNA methyltransferase enzymes

Exam tip

You may be given data from investigations into gene expression to interpret. You may also be given data to evaluate that relates to the relative effects of genetic and environmental factors on phenotype.

are already being used to treat patients with certain kinds of cancer, and an inhibitor of histone deacetylase enzymes is also being used for a specific type of cancer. Further epigenetic drugs are in the course of development.

Summary

- In specialised cells, genes are 'switched on' or 'switched off'.
- Certain genes can be stimulated or inhibited when specific protein transcriptional factors move from the cytoplasm into the nucleus.
- Oestrogen is a steroid hormone that can stimulate transcription of certain genes. It combines with a specific transcriptional factor inside the cell and this complex then enters the nucleus, binds to certain places on the DNA and stimulates transcription.
- Epigenetics involves inherited changes in gene function that do not involve a change in the DNA base sequence.
- Epigenetic changes are caused by changes in the environment.

- They may involve increased addition of methyl groups to cytosine bases in the DNA.
- Alternatively, they may involve the addition of acetyl groups to histone proteins, or the removal of these groups.
- Increased methylation of DNA, or decreased acetylation of histone proteins, inhibit transcription of genes.
- Epigenetic changes are involved in the development of cancer, and in its treatment.
- Another way in which genes are silenced is by RNA interference.
- Short interfering RNA (siRNA) cuts mRNA produced by certain genes so that it cannot be translated.

Gene expression and cancer

A tumour is a group of cells that are dividing too fast. **Benign** tumours are tumours that stay in one place and do not invade other parts of the body. If they can be surgically removed, they usually do not grow back. This does not mean that they are necessarily harmless. For example, a benign tumour in the brain can create pressure on healthy brain tissue and cause considerable harm. **Malignant** tumours are cancerous tumours. If the cancer is not successfully treated at an early stage, the cells spread into other parts of the body. Cells break off from the tumour and spread round the body in the blood or lymph systems, starting secondary tumours, or **metastases**, elsewhere. The process of spreading is called **metastasis**. You can see the stages in the development of cancer in Figure 14.

Cancer occurs when the rate of cell division (mitosis) is uncontrolled. The rate of mitosis is controlled by two groups of genes:

- **Proto-oncogenes**, which control cell division
- **Tumour suppressor genes**, which slow cell division. They also cause programmed cell death (apoptosis) in cells with DNA damage that the cell cannot repair.

A mutated proto-oncogene is called an **oncogene**. This causes a cell to divide too quickly. Proto-oncogenes often code for proteins called **growth factors** that stimulate cell division by binding to complementary receptors in the cell membrane. Other proto-oncogenes code for these protein receptors in the cell membrane. Oncogenes cause cancer by producing too much of these growth factors, or by coding for protein receptors in the cell membrane that stimulate cell division even when no growth factor is present. Mutated tumour suppressor genes are inactivated, so that cells with damaged DNA are not destroyed and the rate of cell division increases. You can see how mutations may result in tumour formation in Figure 15.

Knowledge check 27

Explain why cancer is more likely to be cured if it is detected early.

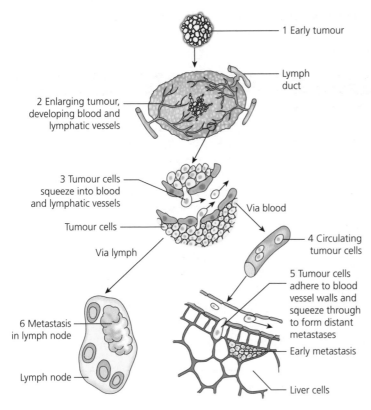

Figure 14 Stages in the development of cancer

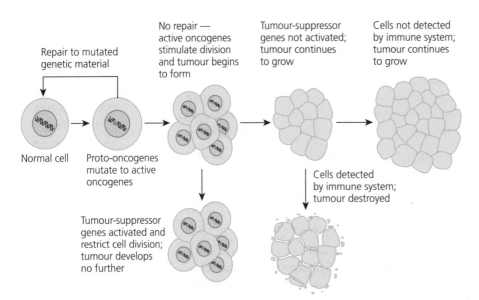

Figure 15 How mutations may result in tumour formation

You have already seen that methylating genes can inactivate them, so hypermethylated tumour suppressor genes will cause cell division to increase, and cells with damaged DNA to survive.

For cancer to develop, several mutations are usually required. These may be gene mutations, epigenetic mutations, or a combination of both.

Some breast cancers are stimulated by the hormone oestrogen. This is because, in these cancers, the cancer cells have a receptor that oestrogen binds to. This then stimulates cell division. People with oestrogen-dependent breast cancers can be treated with drugs that block the oestrogen receptors by binding to them, or by inhibiting the synthesis of oestrogen.

Exam tip

You may be asked to evaluate evidence showing correlations between genetic and environmental factors and various forms of cancer. You may also be asked to interpret information relating to the way in which an understanding of the roles of oncogenes and tumour suppressor genes could be used in the prevention, treatment and cure of cancer.

Summary

- A tumour is a group of cells that is dividing too quickly.
- A benign tumour stays in one place, and usually does not grow back if it is removed.
- Malignant tumours are cancerous tumours.
- Cells can break off from the primary tumour and spread round the body in a process called metastasis, causing secondary tumours called metastases to develop elsewhere.

Using genome projects

The order of bases in the DNA of many organisms, including humans, has now been found as the result of **sequencing projects**. The sequence of bases in the DNA of an organism is called its **genome**. In simpler organisms, once the genome has been found, it is possible to work out the **proteome** (the amino acid sequence of the proteins that the DNA codes for). Once the proteome has been found, it is possible to use this information. For example, scientists have sequenced the genome of the parasite that causes malaria. By sequencing the genome of many different parasites, and searching for differences, it is possible to find the genes that are under the greatest selection pressure. Then the scientists try to find out whether these genes code for proteins that can be used as antigens in vaccine production.

In higher organisms, such as humans, it is much harder to work out the proteome from the genome. This is because the genome contains introns, promoter regions, enhancer regions as well as minisatellite and microsatellite regions between genes. However, once the proteome is determined, this information should be very useful in many ways: for example in the treatment and potential cure of diseases and genetic conditions.

When the first genomes were sequenced, it took a long time for the whole genome to be worked out. However, sequencing methods are now automated and sequences can be obtained quickly.

Knowledge check 28

Explain how comparing the genome sequences of many different malaria parasites can show the genes that are under the greatest selection pressure.

Summary

- The DNA base sequences of many organisms, including humans, have been determined.
- The DNA base sequence of an organism is called its genome.
- The amino acid sequence of the proteins for which the DNA codes is called the proteome.
- In simpler organisms this is easier to work out, and has many applications, including the identification of potential antigens for use in vaccine production.
- In more complex organisms, the proteome is much harder to work out because of the regions of non-coding DNA that are present, such as exons, and regulatory genes.
- Sequencing methods are being improved all the time, and are now automated and much faster than they used to be.

Gene technologies

Recombinant DNA technology

Recombinant DNA technology involves the transfer of a sequence of DNA from one organism, or one species, to another. The organism with the new piece of DNA in its cells is called a **transgenic** organism. The DNA code is universal, so this means that a gene from one species can be transcribed and translated in the same way in a different species.

The fragments of DNA to be transferred can be produced by several methods, including:

- **Converting mRNA to complementary DNA**. Although every cell in an organism has the same DNA, many genes are 'switched off' in specialised cells. For example, a scientist may want to locate the gene for human insulin. This is present in every cell of the body, but only beta cells in the islets of Langerhans in the pancreas are involved in transcribing and translating the gene that codes for insulin. Therefore, a lot of the mRNA in a beta cell from the islets of Langerhans will code for insulin. Figure 16 shows the steps involved.
  - Place the mRNA from the cell with the enzyme **reverse transcriptase** and DNA nucleotides.
  - This enzyme reverses the process of transcription and makes a single-stranded piece of complementary DNA (cDNA).
  - The mRNA is washed out of the mixture.
  - The cDNA is mixed with the enzyme DNA polymerase and free DNA nucleotides.
  - DNA polymerase makes a complementary strand of DNA which converts the cDNA to double-stranded DNA.
- **Using restriction enzymes to cut the gene out of the donor cell**. Restriction enzymes are enzymes that cut DNA whenever they find a specific base sequence, or **restriction site**. Some restriction enzymes make a straight cut across both strands of the DNA (creating 'blunt ends') but many make a staggered cut (leaving overhanging ends called 'sticky ends'). Figure 17 shows the action of a restriction enzyme that leaves sticky ends.

> **Knowledge check 29**
>
> What is the advantage of creating the gene required from mRNA or from a 'gene machine' rather than cutting it out of the donor cell?

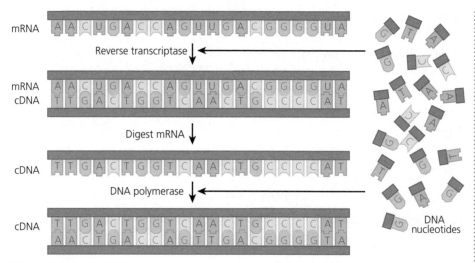

Figure 16 Creating the gene using reverse transcriptase

The scientist chooses a restriction enzyme that will cut the DNA before and after the required gene.

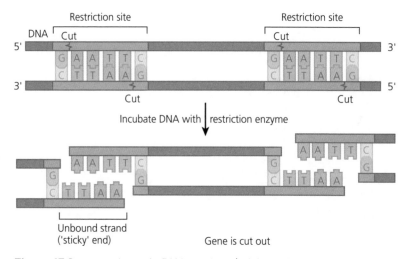

Figure 17 Staggered cuts in DNA produce 'sticky ends'

- **Synthesising the required gene using a 'gene machine'.** If the amino acid sequence in the protein is known, it is possible to use a table of the genetic code to work out the DNA base sequence that will code for the protein. This DNA molecule can be synthesised in an automated computer-controlled machine called a 'gene machine'.

More copies of the gene can be created using the **polymerase chain reaction**. This is an automated technique used to make many copies of a tiny sample of DNA in a very short time. The sample of DNA to be copied is placed in a tube, along with:

- heat-stable DNA polymerase enzyme
- DNA nucleotides
- two primers (short single-stranded pieces of DNA complementary to the 'beginning' and 'end' of the section of DNA to be copied

The tube is placed in a PCR machine, where it undergoes repeated cycles of heating and cooling:

- The DNA is heated to break the hydrogen bonds, making it single-stranded.
- It is cooled so that the primers join to the ends of the single-stranded pieces of DNA.
- It is warmed so that DNA polymerase synthesises a new strand of DNA.

This cycle is repeated again and again. The amount of DNA present doubles with every cycle. This is shown in Figure 18.

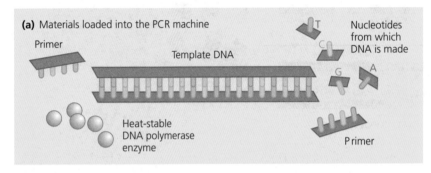

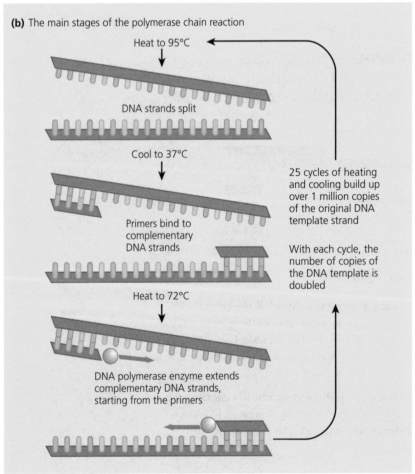

Figure 18 The polymerase chain reaction

The DNA polymerase used in PCR is stable at high temperatures. If it was not thermostable, it would denature when the DNA is heated to 95°C to make the DNA single-stranded, and new enzyme would need to be added every cycle.

Another way of amplifying DNA is to insert the gene into a host cell. Bacteria are often used for this because bacteria contain plasmids, which are very useful.

■ The plasmid is cut open using the same restriction enzyme used to cut the gene from the donor cell. This means the 'sticky ends' of the plasmid and gene will be complementary. If the DNA was synthesised from mRNA or in a 'gene machine', sticky ends are added to its blunt ends.

■ The DNA fragments are incubated with the plasmids and the enzyme **ligase**. Ligase splices the new gene into the plasmid. You can see this in Figure 19.

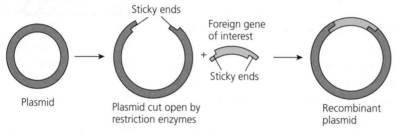

Figure 19 Transferring a gene into a plasmid

Any DNA that has had foreign DNA inserted into it is called **recombinant DNA**, so the plasmid containing the new gene is now called a **recombinant plasmid**. The plasmids are then incubated with bacteria that have been treated so that they are more likely to take up plasmids. However, some bacteria will not take up a plasmid, and others may take up a non-recombinant plasmid. Scientists need to separate the bacteria that contain the recombinant plasmids from the rest. There are two ways of doing this:

■ **Using marker genes**. Marker genes are genes that are inserted into the plasmid as well as the target gene. The marker gene allows the scientist to identify which plasmids contain the target gene. Marker genes code for identifiable features, such as fluorescence or enzymes that make the bacteria antibiotic resistant. One way to use a marker gene is to use a plasmid that already contains genes for tetracycline and ampicillin resistance. A restriction enzyme is used that cuts the plasmid in the middle of the tetracycline resistance gene. The new gene of interest is inserted here. You can see this in Figure 20.

There will be three kinds of bacteria present after the bacteria have been incubated with plasmids.

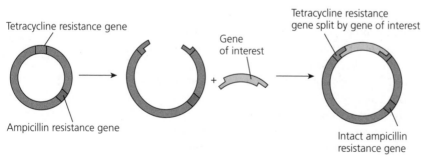

Figure 20 Inserting a gene into a plasmid can split a gene already present

Knowledge check 30

What is the importance of the primer in Figure 18? Why are two different primers needed?

Knowledge check 31

Why is it important that the DNA polymerase enzyme used in PCR is thermostable?

Knowledge check 32

If you started with one molecule of DNA, how many molecules would be present after six cycles?

- Those that have not taken up a plasmid, so these will not grow in any medium containing ampicillin or tetracycline.
- Those that have taken up a non-recombinant plasmid. These will be resistant to both antibiotics and will grow in a medium containing ampicillin or tetracycline.
- Those that have taken up a recombinant plasmid. These are the ones the scientist wants to use. These will be resistant to ampicillin but not to tetracycline. These bacteria will survive on a medium containing ampicillin *only*.
- **Using DNA probes**. A DNA probe is a single-stranded piece of DNA that is complementary to part of the gene of interest. It is made fluorescent or radioactive, so that it can be detected later.
  - The bacteria are cultured on agar gel in Petri dishes. They are diluted, so that individual cells are separated on the agar surface. The bacteria are incubated, so that they divide to form colonies.
  - A piece of sterile filter paper is 'blotted' onto the Petri dish so that a few bacteria in each colony transfer to the filter paper.
  - The cells from the filter paper are broken open to expose their DNA and the DNA is treated to make it single-stranded.
  - A radioactive probe is added to the filter paper. This will only bind to complementary DNA.
  - Excess probe is washed off and the positions where the radioactive DNA has bound are found by placing the filter paper against X-ray film. The radioactive probe causes 'fogging' on the X-ray film.
  - The positions where fogging appears on the X ray film correspond to the colonies of bacteria on the Petri dish that contain the gene of interest. The scientists can culture these bacteria, either to clone the gene or to produce the protein that the gene codes for.

This process is shown in Figure 21.

When a gene of interest is transferred to a new cell, a promoter and terminator sequence need to be added as well, to ensure that the gene is transcribed properly. A plasmid is just one example of a **vector**. Another common example of a vector is a virus. Vectors can be used to transform eukaryotic cells as well as bacteria. For example, crops may be transformed so that they are resistant to weedkiller or resistant to insect pests.

There are issues relating to these uses of gene technology. If a vector such as a virus is used, we cannot be certain exactly where the new gene will insert itself into the host cell's genome. There is a chance that it will disrupt another important gene, or 'switch on' a gene that has been 'switched off'.

There are also concerns that some genetically modified organisms create profits for large corporations but have little benefit to other people. For example, the large American company Monsanto produces soya plants that are resistant to the weedkiller, glyphosate. They claim that this is an advantage because the farmer can apply glyphosate to the crop, and this will kill all the weeds but not the crop. They claim that less weedkiller will be used in growing this crop, than is used for conventional crops, as all the weeds can be destroyed early in the season. However, Monsanto not only make profits from seeds for the glyphosate-resistant soya, but they also produce glyphosate. Some people are concerned that this increases profits for wealthy American farmers and makes it harder for poor farmers in the developing world to compete.

Knowledge check 33
All the bacteria in a colony are genetically identical. Explain why.

Knowledge check 34
Why is it important to wash off the excess probe?

Knowledge check 35
PCR is called an 'in vitro' method of copying a gene. If bacterial cells are used, this is an 'in vivo' method. What do these two terms 'in vitro' and 'in vivo' mean?

Exam tip
Avoid unscientific arguments such as 'this is unnatural' or 'we must not play God'. Express views in a balanced way and explain the basis for the viewpoint.

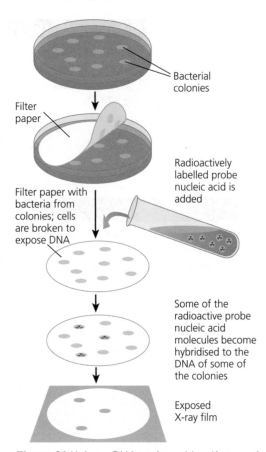

Bacterial colonies

Filter paper

Filter paper with bacteria from colonies; cells are broken to expose DNA

Radioactively labelled probe nucleic acid is added

Some of the radioactive probe nucleic acid molecules become hybridised to the DNA of some of the colonies

Exposed X-ray film

Figure 21 Using a DNA probe to identify transformed bacteria

Some people are concerned that the additional genes might pass to other species — for example, weeds might become weedkiller resistant — via cross-breeding. However, this is just as likely to happen if traditional breeding methods are used and all genetically modified organisms have to be extensively tested in field trials before they are approved.

Another viewpoint is that we should not interfere with the genetics of particular species. However, over the last few millennia, humans have been selectively breeding plants and animals, so our farm crops and animals are now genetically very different from the wild organisms they developed from.

On the other hand, many people feel that genetically modified organisms are the way forward if we wish to produce enough food for a growing human population. They argue that genetic modification is simply doing what traditional breeding methods have done, but more quickly. We can use these techniques to produce crops that are drought resistant, disease and pest resistant, so that crop yields are greater.

In addition, we can produce useful medicines like human insulin, factor VIII and human growth hormone from genetically modified bacteria. These can be used to treat people with serious health conditions. People have expressed fears that gene technology might lead to the development of 'designer' babies. Most doctors would find this completely unacceptable, and genetically modifying humans has been made illegal by international agreements.

Exam tip

You do not need to know any specific examples of genetic modification, but you will be expected to evaluate the social, ethical and financial aspects of any example that is presented to you. Do not try learning a list of issues, but have the confidence to relate to the example the examiner gives you.

Gene therapy is a method of treating genes that cause disease. Examples include cystic fibrosis and sickle cell anaemia. These genes only cause problems in certain cells of the body so scientists are trying to find ways of 'correcting' these genes. In the future, it may be possible to silence a harmful gene, but current technology involves efforts to insert copies of the 'healthy' allele into affected tissues. For example, cystic fibrosis has its most harmful effects in the cells lining the respiratory tract. In trials, a virus has been modified to transfer the 'healthy' allele into these cells. It is administered using an inhaler, similar to those used by asthmatics. So far, the trials have had limited success as many cells fail to take up the new gene and the body's immune system has attacked the virus particles.

Knowledge check 36

The inhaler treatment has to be repeated at regular intervals, e.g. every 2 to 3 months. Explain why.

Summary

- Recombinant DNA tchnology involves transferring DNA from one organism, or one species, into another.
- The fragment of DNA to be transferred can be produced by converting mRNA to cDNA using reverse transcriptase; using restriction enzymes to cut it out of DNA; or by synthesising the gene in a 'gene machine'.
- Fragments of DNA may be copied using the polymerase chain reaction or in plasmids in bacterial cells.
- Transferred genes need to have promoter and terminator regions added so that they can be transcribed correctly in the new host cell.
- Plasmids may be cut open with restriction enzymes so that the new gene can be added.

- The gene is inserted into the plasmid using the enzyme ligase.
- The plasmid containing the new gene is a recombinant plasmid.
- Host cells that take up the plasmid vector are said to be transformed.
- Marker genes are genes that are transferred along with the gene of interest, and they allow host cells containing the transferred gene to be identified.
- There are social, ethical and financial issues associated with the use of recombinant DNA technology in agriculture, industry and medicine.
- Gene therapy involves transferring a 'healthy' copy of a gene into a host cell to correct a genetic defect.

Differences in DNA between individuals of the same species can be exploited for identification and diagnosis of heritable conditions

Gene probes may be used to screen people for specific alleles of genes. The DNA of a person is cut into fragments using a restriction enzyme, then the fragments of DNA are separated using **gel electrophoresis**. This is shown in Figure 22.

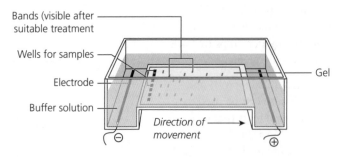

Figure 22 Separating DNA using gel electrophoresis

The mixture of DNA fragments is loaded into a well in an agarose gel. This is covered with a buffer solution and an electric current applied. The pieces of DNA move towards the positive electrode, and the smaller fragments move faster than the longer fragments. The DNA fragments are transferred to a nylon membrane, then they are treated to make them single-stranded. After this, gene probes are applied. These bind to any complementary sequences they find. This is called **DNA hybridisation**. The membrane is washed to remove excess probe. The membrane is then examined to determine whether the probe has bound. If the probe is radioactive, this is done using X-ray film, but if the probe is fluorescent, the membrane will be examined under UV light. If the probe has bound to the DNA, this means that this person's DNA does contain the allele being tested for.

Gene technology can also be used to produce DNA microarrays (DNA chips), which contain thousands of gene probes, each for a specific gene. DNA from a person being tested is split into fragments and added to the microarray. If a probe binds to the person's DNA a colour change occurs. This can be used to test a person's genome for a whole range of genes.

Information from DNA analysis using gene probes can be used by genetic counsellors to inform people who are at risk of developing a genetic condition, and/or who may be at risk of having a child with a genetic condition. Increasingly, this technique is used to identify specific mutations that have caused a person's cancer. This means that an appropriate therapy, specific to that mutation, may be used to treat the person's cancer. This is an example of **personalised medicine**.

> ### Knowledge check 37
> Why does DNA move towards the positive electrode?

Exam tip

There are ethical issues involved in using gene probes to identify specific alleles in a person's DNA. For example, this may lead to a couple at risk of having a child with a genetic defect undergoing IVF treatment and the selection of 'healthy' embryos. Some people think that the creation of embryos that will not be allowed to develop is wrong. Also, when one family member is tested for the presence of an allele, this may also give information about another family member's genome. You may be required to evaluate the use of gene probes in a specific situation presented by the examiner.

Summary

- Labelled gene probes may be used to locate specific alleles in a person's DNA.
- This allows scientists to detect alleles that may pose health risks to a person, or a response to a drug.
- For example, a person with a specific mutation that causes cancer may respond well to a particular drug that has no effect at all if another mutation is present.

- A person's DNA is analysed by cutting it into fragments using a restriction enzyme, separating the fragments using gel electrophoresis, and transferring the DNA to a nylon membrane.
- The DNA is made single-stranded and then the probe is added.
- If the probe binds, then the allele being tested for is present.
- This is useful in personalised medicine and in genetic counselling.

Genetic fingerprinting

Genetic fingerprinting is a technique that examines the non-coding DNA of different people. In between the genes there are repetitive sequences of DNA bases called tandem repeats. The number of these tandem repeats varies considerably between people. These are therefore called **variable number tandem repeats (VNTRs)**. Although VNTRs vary between individuals, the number of VNTRs in an individual is inherited from her or his parents. Therefore, people with a similar pattern of VNTRs are likely to be related. A sample that contains as little as one molecule of DNA can be analysed, since the polymerase chain reaction can be used to produce many copies of the DNA. Genetic fingerprints are prepared as shown in Figure 23.

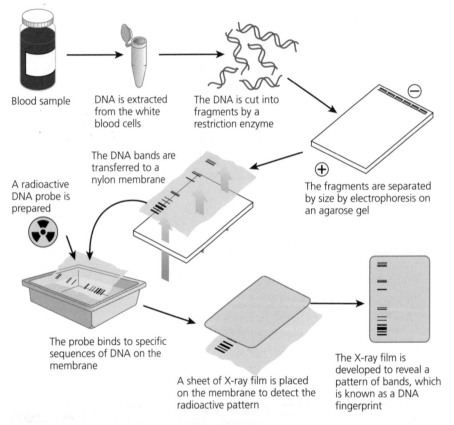

Figure 23 The main stages in obtaining a genetic fingerprint

The chance of two individuals having the same pattern of VNTRs (unless they are identical twins) is very low. Therefore, genetic fingerprints can be used in forensic science, to identify a criminal. For example, if a person has broken into a building or murdered somebody, it is extremely likely that they will leave a trace of their DNA at the crime scene: for example in a drop of blood. This sample can be analysed to find the genetic fingerprint of the burglar or murderer. Furthermore, this allows innocent suspects to be eliminated. In the case of rape, a vaginal swab can be taken and analysed. All the bands in the resulting fingerprint that do not come from the victim

Knowledge check 38

In forensic science DNA may be contaminated. For example, if the police find a human body part it is likely to be heavily contaminated with DNA from microorganisms. Suggest how the use of PCR enables forensic scientists to identify DNA from the human rather than contaminating DNA.

must have come from the rapist. Another use of genetic fingerprinting is in identifying bodies, or body parts. The genetic fingerprint can be compared to living relatives.

Genetic fingerprinting can also be used to show genetic relationships. All the bands in a child's DNA fingerprint that do not come from the mother must come from the father. This can be used to prove or disprove paternity, for example.

Genetic fingerprinting is used in animal and plant breeding. It can be used to find individuals that are as unrelated as possible, so they can be interbred to increase genetic diversity in a captive population.

Exam tip

Genetic fingerprinting is the term for analysis of VNTRs. This analyses non-coding DNA, but does not give information about the individual's phenotype or medical conditions. However, the term is sometimes used in the general sense of 'analysing DNA'. Analysing DNA using gene probes, or sequencing the genome, can be useful in medical diagnosis as you have already seen. It can also be used to analyse plants. For example, one species of plant may have a gene that codes for a chemical that is medically useful but produced in tiny amounts. Using a gene probe, scientists may study closely related species to see whether they produce the same chemical but in larger concentrations.

Summary

- In the genome of an organism there are many VNTRs (variable number tandem repeats).
- These are repetitive sequences of bases in the non-coding regions of the DNA.
- The VNTRs of an individual are inherited from parents, so the chances of two unrelated individuals having the same pattern of VNTRs is extremely low.
- Samples of DNA for analysis can be amplified using the polymerase chain reaction.
- The DNA is cut into fragments using a restriction enzyme, separated using gel electrophoresis, transferred to a nylon membrane.
- It is made single-stranded and then a radioactive probe is added.
- The probe binds to the VNTRs.
- The pattern of bands, resembling a barcode, becomes visible on X-ray film.
- This technique can be used to identify genetic relationships and also to assess the genetic variability of a population.
- In forensic science, it can be used to identify criminals or murder victims.

Questions & Answers

For A-level biology, your exams will be structured as follows:

Paper 1	Paper 2	Paper 3
Any content from topics 1–4 including relevant practical skills	Any content from topics 5–8 including relevant practical skills	Any content from topics 1–8 including relevant practical skills
Written exam 2 hours 91 marks worth 35% of A-level	Written exam 2 hours 91 marks worth 35% of A-level	Written exam 2 hours 78 marks worth 30% of A-level
76 marks: mixture of long and short answer questions 15 marks: extended response	76 marks: mixture of long and short answer questions 15 marks: comprehension	38 marks: structured questions including practical techniques 15 marks: critical analysis of experimental data 25 marks: essay from a choice of 2 titles

The topics in this book are examined in paper 2 (together with topics 5 and 6) and in paper 3 (together with topics 1–4 and 5 and 6). Paper 3-type questions are included here, but they are focused mainly on topics 7 and 8 as those are the focus of this guide.

There are several ways of using this section. You could:

- 'hide' the answers to each question and try the question yourself. It need not be a memory test — use your notes to see if you can actually make all the points you ought to make
- check your answers against the students' responses and make an estimate of the likely standard of your response to each question
- check your answers against the comments to see where you might have failed to gain marks
- check your answers against the terms used in the question — for example, did you *explain* when you were asked to, or did you merely *describe*?

All student responses are followed by detailed comments. These are preceded by the icon ⓔ and indicate where credit is due. In the weaker answers, they also point out areas for improvement, specific problems, and common errors such as lack of clarity, weak or non-existent development, irrelevance, misinterpretation of the question and mistaken meanings of terms.

Tips for answering questions

Use the mark allocation. Generally, 1 mark is allocated for one fact, concept or item in an explanation. Make sure your answer reflects the number of marks available.

Respond appropriately to the command words in each question, i.e. the verb the examiner uses. The terms most commonly used are explained below.

- **Describe** — this means 'tell me about...' or, sometimes, 'turn the pattern shown in the diagram/graph/table into words'; you should not give an explanation.
- **Explain** — give biological reasons for *why* or *how* something is happening.

- **Calculate** — add, subtract, multiply, divide (do some kind of sum!) and show how you got your answer — *always* show your working!
- **Compare** — give similarities *and* differences between…
- **Complete** — add to a diagram, graph, flowchart or table.
- **Name** — give the name of a structure/molecule/organism etc.
- **Suggest** — give a plausible biological explanation for something; this term is often used when testing understanding of concepts in an unfamiliar context.
- **Use** — you must find and include in your answer relevant information from the passage/diagram/graph/table or other form of data.

General advice on writing a good essay

The essay is marked out of 25; 16 marks are for scientific knowledge, 3 for breadth, 3 for relevance, and 3 for quality of language.

- To get the breadth marks, use examples from as many different modules as possible.
- Quality of language refers mainly to good use of scientific terminology. Avoid words like 'signals' or 'messages' when you mean 'nerve impulses', and remember that enzymes become denatured, not 'killed'.
- Make sure everything you include is relevant to the title.
- Write five good, detailed paragraphs. Do not waste time with an introduction or conclusion.
- To keep you on track, it's a good idea to start each paragraph with the essay title, e.g. 'Another way in which inorganic ions are important in biology is…'.
- To get 16 marks for scientific content you need to include relevant detailed information that is beyond A-level. However, be careful — sometimes students are so eager to get these marks that they include information that is not relevant (see student B below) or information that is not really beyond A-level, such as an example from a television programme. These can actually lose marks.

A note on statistics

The examiners will expect you to be familiar with the use of statistical tests and standard deviation. It is unlikely that you will be asked to carry out a complete statistical analysis in the exam, but you may well be asked to show understanding of the tests, e.g. by suggesting a suitable test; writing a suitable null hypothesis; looking up a value on a table and interpreting its significance; and interpreting what a given value of p means, or calculating the number of degrees of freedom that apply.

You will be expected to be familiar with correlation coefficients (see Spearman's rank correlation coefficient in the practical on pp. 74–75), the chi-squared test (see knowledge check 18 on p. 24) and the t-test. An example of the t-test is shown below.

The t-test is used to find out whether the means of two samples are significantly different. For example, a student measured the length of the index fingers of male and female students in her class.

The null hypothesis is: There is no difference in the length of the index finger in males and females.

The formula for the t-test is:

$$t = \frac{\bar{x}_1 - \bar{x}_2}{\sqrt{\dfrac{S_1^2}{N_1} + \dfrac{S_2^2}{N_2}}}$$

where $\bar{x}_1$ is the mean of first data set

$\bar{x}_2$ is the mean of second data set

S_1 is the variance of sample 1, $\dfrac{\Sigma(x_1 - \bar{x}_1)^2}{n_1}$

S_2 is the variance of sample 2, $\dfrac{\Sigma(x_2 - \bar{x}_2)^2}{n_2}$

N_1 is the number of measurements in the first data set

N_2 is the number of measurements in the second data set

Females			Males		
x_1	$x_1 - \bar{x}_1$	$(x_1 - \bar{x}_1)^2$	x_2	$x_2 - \bar{x}_2$	$(x_2 - \bar{x}_2)^2$
77	2.2	4.84	84	4.2	17.64
75	4.2	17.64	91	2.8	7.84
81	1.8	3.24	90	1.8	3.24
83	3.8	14.44	87	1.2	1.44
80	0.8	0.64	89	0.8	0.64
$\bar{x}_1 = 79.2$		$\Sigma = 40.80$	$\bar{x}_2 = 88.2$		$\Sigma = 30.8$

$$S_1 = \frac{40.8}{5} = 8.16 \qquad S_2 = \frac{30.8}{5} = 6.16$$

$$t = \frac{\bar{x}_1 - \bar{x}_2}{\sqrt{\dfrac{S_1^2}{N_1} + \dfrac{S_2^2}{N_2}}}$$

$$t = \frac{79.2 - 88.2}{\sqrt{\left(\dfrac{8.16^2}{5} + \dfrac{6.16^2}{5}\right)}}$$

$$= \frac{9}{\sqrt{13.317 + 7.589}}$$

$$= \frac{9}{\sqrt{20.906}}$$

$$= \frac{9}{4.572}$$

$$= 1.97$$

Degrees of freedom $= (n_1 + n_2) - 1 = 9$

This value has to be looked up on a t-table:

Degrees of freedom	Significance level					
	20% (0.02)	10% (0.01)	5% (0.05)	2.5% (0.025)	1% (0.01)	0.01% (0.001)
8	0.889	1.397	1.860	2.308	2.896	4.501
9	0.883	1.383	1.883	2.262	2.821	4.297
10	0.879	1.372	1.812	2.228	2.764	4.144

For $p = 0.05$ and 9 degrees of freedom the critical value is 1.883.

The calculated value of t is 1.97. This is higher than the critical value, so the student must reject the null hypothesis. Therefore there is a significant difference between the mean length of index finger for males and females. The probability of the difference being due to chance is less than 0.05 or 5%.

■ Paper 2-type questions

Question 1

A piece of DNA was cut into fragments using an enzyme. The fragments were then separated using gel electrophoresis.

(a) Name the type of enzyme used to cut the DNA into fragments. (AO1)   (1 mark)

Student A

(a) Restriction endonuclease

ⓔ This is a correct answer and scores 1 mark.

Student B

(a) Restriction enzyme

ⓔ This also gets a mark for a correct answer. Note that you can call the enzyme a restriction endonuclease or simply a restriction enzyme.

(b) (i) The table shows the number of base-pairs present in each of the fragments.

Fragment	Number of base-pairs ($\times 10^3$)
1	2.71
2	4.56
3	5.46
4	5.97
5	6.73
6	12.23

The diagram shows the electrophoresis gel after electrophoresis had taken place. The bands show the position of the different DNA fragments.

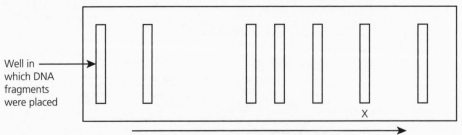

Well in which DNA fragments were placed

X

Direction in which DNA fragments travelled

The enzyme cut the DNA at a particular sequence of bases. How many times did this base sequence appear in the original piece of DNA? Explain your answer. (AO1/2)   (2 marks)

Student A

(b) (i) Five times because six fragments are produced.

ⓔ 2 marks here, for the correct number and a correct reason.

Student B

(b) (i) Six times because it cut the DNA into six pieces.

ⓔ The answer is wrong so does not get a mark, but the explanation is correct so that scores 1 mark.

(b) (ii) How many base-pairs are there in the fragment marked X on the diagram? Explain how you arrived at your answer. (AO2)   (2 marks)

Student A

(b) (ii) 4.56×10^3 because the smallest fragments move furthest, and this band is the second shortest.

ⓔ 2 marks here for a fully correct answer.

Student B

(b) (ii) 6.73 because band X is the second longest

ⓔ No marks here because this student has forgotten that the longer bands move more slowly. Also, the student should have put this length as 6.73×10^3, though this mistake alone is unlikely to lose the mark.

(c) DNA fragments move towards the positive electrode. Explain why. (AO1)   (1 mark)

Student A

(c) Each nucleotide contains a negatively charged phosphate group.

ⓔ This is correct and gains the mark.

Student B

(c) The phosphate groups in the DNA are negatively charged.

ⓔ This also gets the mark for being correct.

Question 2

(a) Complete the table with the names of the missing enzymes. (AO1)   (5 marks)

Type of enzyme	Function
	Breaks hydrogen bonds between complementary bases during DNA replication
	Makes a single-stranded piece of DNA from a piece of RNA
	Joins DNA nucleotides together during DNA replication
	Cuts DNA at a specific base sequence
	Joins pieces of DNA together

Questions & Answers

Student A

(a)

Type of enzyme	Function
DNA helicase	Breaks hydrogen bonds between complementary bases during DNA replication
Reverse transcriptase	Makes a single-stranded piece of DNA from a piece of RNA
DNA polymerase	Joins DNA nucleotides together during DNA replication
Restriction enzyme	Cuts DNA at a specific base sequence
DNA ligase	Joins pieces of DNA together

e 5 marks here for a fully correct answer

Student B

(a)

Type of enzyme	Function
Hydrolase	Breaks hydrogen bonds between complementary bases during DNA replication
DNA polymerase	Makes a single-stranded piece of DNA from a piece of RNA
DNA ligase	Joins DNA nucleotides together during DNA replication
Restriction enzyme	Cuts DNA at a specific base sequence
DNA polymerase	Joins pieces of DNA together

e 1 mark for restriction enzyme. Breaking the hydrogen bonds in DNA is not a hydrolysis; and DNA polymerase joins individual nucleotides together, not pieces of DNA.

(b) Pig farms produce a lot of manure. Within the manure is a lot of phytase, a phosphorus-containing compound. Pigs cannot digest phytase so it can cause eutrophication if it leaches into ditches and streams. Scientists have produced the Enviropig using genetic engineering. This pig digests phytase. Therefore the manure from Enviropigs is less likely to cause eutrophication.

   (i) What is eutrophication? (AO1)   (4 marks)

   (ii) Give the role of the following in producing the Enviropig

       Vector (AO1)   (1 mark)

       Marker gene (AO1)   (2 marks)

Student A

(b) (i) Nitrates and phosphates enter the stream and increase the growth of surface algae ✓. This blocks out the light from plants further down ✓. These plants cannot photosynthesise, so they die ✓. bacteria respire feeding on the dead plants ✓. They use up the oxygen in respiration, so fish and other organisms die.

🅔 This full answer gains full marks and would have gained a fifth had a further mark been available.

> (ii) A vector is something like a plasmid or virus that carries the new piece of DNA into the host cell.
>
> A marker gene is a gene that is inserted along with the target gene. It is for an identifiable feature, e.g. fluorescence, so that the cells containing the target gene can be identified.

🅔 Full marks here for a clear function of both a vector and marker gene.

Student B

> (b) (i) The manure causes algae to grow. The plants lower down can't get enough light for photosynthesis ✓, so they die and bacteria decompose them ✓. They use up oxygen in their respiration ✓, so fish and other organisms die ✓.

🅔 This answer does not get the first mark because the growth of surface algae is not well expressed. However, the student gives a clear account of the rest of the story and gains a full 4 marks.

> (ii) A vector is used to insert the new gene into the cell. A maker gene is inserted into the cell at the same time as the gene to digest phytase.

🅔 1 mark for a correct explanation and 1 mark for marker gene. There is no second mark for the marker gene answer because the student has not explained that this allows identification of the cells that have been modified successfully.

(c) **Do you think that farmers should be encouraged to farm the Enviropig, rather than normal pigs? Give reasons for your answer. (AO3)** (5 marks)

Student A

> (c) Yes they should because this pig causes less pollution as there will be less eutrophication ✓. Also, the pig can digest phytase, so it will utilise more of its food, giving a greater growth rate ✓. However, many people are worried about genetic engineering and so the farmer may not be able to sell the pigs ✓. The gene may interfere with other genes in the pig and cause harm to the pig or cause the pig to produce a toxic compound ✓, though this is unlikely. The farmer will not be able to sell his meat as organic ✓, although having pigs that digest more of their food may reduce the cost of meat production for the farmer.

🅔 This student makes 5 valid points and gets full marks. Note that the student needs to refer to points both for and against to gain full marks.

Questions & Answers

ⓔ This gets only 1 mark. Although the student understands the pig will cause less pollution, this is unspecific. The student does not explain why the pig will be more profitable. The argument against genetic engineering is not made scientifically. The student should explain a concern about genetic engineering in a scientific way (as student A did) and/or explain the concerns that consumers have which might affect the farmer's ability to sell his livestock.

Question 3

Phlox is a type of garden flower. Flower shape is controlled by two codominant alleles. Plants homozygous for S^1 have broad flowers. Plants homozygous for S^2 have 'cuspidata' flowers, and heterozygous plants have 'fimbriata' flowers. Another gene controls whether the petals lie flat or form a funnel shape. The allele *F* for flat petals is dominant over the allele *f* for funnel-shaped petals.

(a) What are *codominant* alleles? (AO1)   (1 mark)

Student A

(a) These are alleles that produce a different phenotype in a heterozygote from both homozygotes.

ⓔ A fully correct answer gaining the mark.

Student B

(a) This is when one allele is not dominant over the other.

ⓔ This gains no marks. This tells you one allele is not dominant over the other, but does not explain what it is. It can be difficult to explain the meaning of genetic terms clearly, so practise this before the exam.

(b) Two plants, both with fimbriata flowers and heterozygous for petal shape, were crossed. Use a genetic diagram to show the ratio of phenotypes and genotypes expected in the offspring. (AO2)   (4 marks)

Student A

(b) Parent phenotypes   fimbriata, flat   ×   fimbriata, flat

Parent genotypes S^1S^2Ff × S^1S^2Ff ✓

Gametes S^1F S^1f S^2F S^2f × S^1F S^1f S^2F S^2f ✓

	S^1F	S^1f	S^2F	S^2f
S^1F	S^1S^1FF	S^1S^1Ff	S^1S^2FF	S^1S^2Ff
S^1f	S^1S^1Ff	S^1S^1ff	S^1S^2Ff	S^1S^2ff
S^2F	S^1S^2FF	S^1S^2Ff	S^2S^2FF	S^2S^2Ff
S^2f	S^1S^2Ff	S^1S^2ff	S^2S^2Ff	S^2S^2ff

3 broad, flat   1 broad, funnel

3 cuspidata, flat   1 cuspidata, funnel   ✓✓

6 fimbriata, flat   2 fimbriata, funnel

@ Full marks here for a fully correct answer that is set out well.

Student B

(b)

	S^1F	S^1f	S^2F	S^2f
S^1F	S^1S^1FF	S^1S^1Ff	S^1S^2FF	S^1S^2Ff
S^1f	S^1S^1Ff	S^1S^1ff	S^1S^2Ff	S^1S^2ff
S^2F	S^1S^2FF	S^1S^2Ff	S^2S^2FF	S^2S^2Ff
S^2f	S^1S^2Ff	S^1S^2ff	S^2S^2Ff	S^2S^2ff

3 broad, flat   1 broad, funnel

3 cuspidata, flat   1 cuspidata, funnel   ✓✓

6 fimbriata, flat   2 fimbriata, funnel

@ This student gets only 3 marks (including 1 mark for the table). Although the right answer is given, this student has not written down the parental genotypes and this was expected. The gametes can be deduced from the Punnett square.

Question 4

The Bali starling *(Leucopsar rothschildi)* is a bird found only on the island of Bali. It is critically endangered because of loss of habitat and trapping for sale as songbirds. In 2001 the world population of Bali starlings fell to only 6 individuals. Captive breeding in zoos has increased the population, and some captive-bred birds have been released back into the wild.

(a) Zoos regularly exchange birds with each other for their captive breeding programmes. Explain why. (AO2)   (2 marks)

Student A

(a) This increases genetic diversity as related birds will have similar alleles ✓. If the population has low genetic diversity the zoo population is more susceptible to environmental change, such as a disease ✓.

@ 2 marks here for a good answer with explanation.

Questions & Answers

> **Student B**
>
> **(a)** This increases genetic diversity by bringing in new alleles from unrelated birds ✓.

e Just 1 mark here. This student does not explain why increased genetic diversity is an advantage.

(b) Name the taxon represented by *Leucopsar*. (AO1) (1 mark)

> **Student A**
>
> **(b)** Genus ✓

e 1 mark for the correct answer.

> **Student B**
>
> **(b)** Species

e No mark here. The student is confused because we say the species is *Leucopsar rothschildi*, but Leucopsar is the genus, and rothschildi is the species.

(c) Suggest how DNA analysis may be useful to zoos participating in the captive breeding of the Bali starling. (AO2) (2 marks)

> **Student A**
>
> **(c)** They can sequence the genome of different birds ✓ and interbreed birds with the greatest numbers of differences ✓.

e A good answer, well expressed, worth 2 marks.

> **Student B**
>
> **(c)** They can choose birds that are genetically different to breed together ✓.

e Just 1 mark here, as this student has not explained how you would identify birds that are genetically different.

Question 5

In mice, the Fas gene causes cell death. In some kinds of liver disease, the Fas gene leads to liver damage because it destroys damaged cells. In one study, scientists injected specific siRNA that targeted the Fas gene into mice with hepatitis. This reduced liver cell destruction and allowed most of the mice to recover. Scientists now hope that Fas siRNA may be useful in the treatment of acute and chronic liver disease in humans.

(a) Describe how siRNA reduced liver damage in mice. (AO1)   (4 marks)

> **Student A**
>
> **(a)** An enzyme called dicer attaches to the siRNA ✓. It separates into two strands and one strand of the siRNA binds to the mRNA transcribed from the Fas gene ✓. RISC attaches to the siRNA/mRNA complex ✓. The mRNA is cut into smaller pieces, so the Fas protein is not made ✓.

ⓔ 4 marks here for a fully correct answer.

> **Student B**
>
> **(a)** The antisense strand of the siRNA binds to the mRNA made by the Fas gene ✓. The mRNA is cut into smaller pieces so the mRNA can't be translated ✓.

ⓔ Only 2 marks here. Although this student knows how siRNA works, some key details are omitted.

(b) (i) Explain how the siRNA was specific to the Fas gene. (AO2)   (2 marks)

> **Student A**
>
> **(b) (i)** The base sequence on the siRNA was complementary ✓ to part of the mRNA produced by the Fas gene ✓.

ⓔ 2 marks for a fully correct answer.

> **Student B**
>
> **(b) (i)** The base sequence of the siRNA was complementary ✓ to the Fas gene.

ⓔ Just 1 mark here. The siRNA is complementary to the Fas gene, but it works by binding to the mRNA so the second mark is not awarded.

(b) (ii) Some people thought that this study was unethical. Do you agree? Give reasons for your answer. (AO3)   (4 marks)

> **Student A**
>
> **(b) (ii)** It could be seen as unethical as the rats were given hepatitis, and they were made ill ✓. Many people think it is unethical to carry out experiments on animals as they cannot give consent ✓. On the other hand, this would be much more unethical if it was tested on humans without testing on animals first ✓, and if the scientists can get this to work properly it could cure a lot of disease and save human lives ✓.

ⓔ 4 marks for four sensible points, and both sides of the argument are covered.

Questions & Answers

Student B

(b) (ii) Some people think it is unfair to experiment on animals because it can cause them to suffer ✓. Other people think you must test treatments on animals before trying them on humans, as you could cause humans to die if you tested it on them first ✓.

ⓔ 2 marks here for two valid points. However, this student has not developed her/his argument as fully as student A, so cannot gain full marks.

Paper 3-type questions

Question 1

Figure 1 shows the inheritance of glucose-6-phosphate dehydrogenase (G6PD) deficiency in one family from Afghanistan.

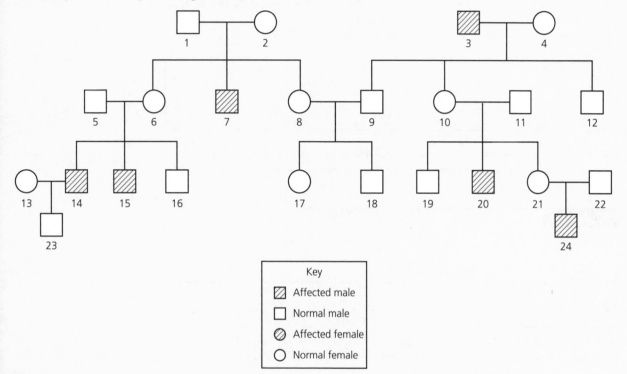

Key

- ▧ Affected male
- ▢ Normal male
- ⊘ Affected female
- ○ Normal female

Figure 1

(a) What is the evidence that this allele is:
  (i) recessive (AO1)   (1 mark)
  (ii) sex-linked (AO1)   (1 mark)

(b) Give the possible genotype(s) of:
  (i) individual 17 (AO1)   (1 mark)
  (ii) individual 21 (AO1)   (1 mark)

(c) In this Afghan population, 10% of males were found to be G6PD deficient. What percentage of females would you expect to be G6PD deficient in this population? Explain your answer. (AO2)   (3 marks)

(d) G6PD is a serious condition which can result in severe anaemia and jaundice. There is no known cure. However, people who are G6PD deficient have some resistance to malaria. Malaria is a potentially fatal disease that was, until recently, common in parts of the Middle East and around the Mediterranean. Use this information to suggest why G6PD deficiency is much commoner in Mediterranean and Middle East populations than in northern Europeans. (AO2)   (3 marks)

Student A

(a) (i) 1 and 2 don't have G6PD deficiency, but 7 does.

(ii) Six males are affected and no females

ⓔ This student gains a mark in (i). It is important to use the numbers of individuals being referred to. Parents 10 and 11 with son 20, or 21 and 22 with son 24, or 5 and 6 with sons 14 and 15 are alternative correct answers. For part (ii) this is a valid and correct answer, especially as there are several affected males but no females.

Student B

(a) (i) Most of the people don't have it

(ii) Only males have it

ⓔ This student does not get a mark in (a) (i). Recessive does not mean 'rare'; there are many recessive conditions that are very common in populations. In (ii) the student does get a mark.

Student A

(b) (i) $X^G X^g$ or $X^G X^G$

(ii) $X^G X^g$

ⓔ Student A gets both marks here. The student realises that there are two possible genotypes for individual 17, as there is no firm evidence to decide between the homozygous and heterozygous state. However, 21 must be heterozygous as she has a son with G6PD deficiency.

Student B

(b) (i) NN

(ii) Nn

ⓔ This student has forgotten that this is a sex-linked condition, so the nomenclature is wrong. Also the student has not realised that there are two possible answers for individual 17. The student has understood that individual 21 is heterozygous, but has not recognised that the condition is sex-linked.

Student A

(c) Men have only one X chromosome, so if 10% of the males in the population have G6PD deficiency then 10% of the X chromosomes in the population must carry this allele. Therefore there is a 0.1 chance of a person inheriting an X chromosome with this allele in this population. Women inherit two X chromosomes, so the chance of inheriting two X chromosomes with G6PD deficiency is $0.1 \times 0.1 = 0.01$. Therefore 1% of the women in this population would be expected to have G6PD deficiency.

ⓔ This is an excellent answer and gets full marks. This is a difficult question, but this student clearly understands the principles involved.

> **Student B**
>
> **(c)** G6PD deficiency is homozygous recessive. Therefore $q^2 = 0.1$,
> $q = \sqrt{0.1} = 0.316$
>
> $p = 1 - q = 1 - 0.316 = 0.684$

ⓔ Student B has tried to use the Hardy–Weinberg formula, but this is not appropriate for a sex-linked gene. The student needed to think through another way to find the frequency of the sex-linked allele. No marks here for student B.

> **Student A**
>
> **(d)** Although G6PD deficiency is a disadvantage to health, in areas where malaria is common people with this condition are more likely to survive ✓. They reproduce and pass on their allele ✓. This increases the frequency of the G6PD allele in the population ✓.

ⓔ This is well expressed. The student has made it clear that there is an advantage in having the G6PD allele in these areas and so it is more likely to be passed on to offspring. Therefore the allele increases in frequency.

> **Student B**
>
> **(d)** The people most likely to die of malaria in these areas would be those who don't have G6PD ✓. Therefore the normal allele will not be passed on to offspring as often as the G6PD allele ✓, so the normal allele will decrease in frequency ✓.

ⓔ This gets full marks, too, but this student has expressed the answer in terms of the normal allele decreasing in frequency, which is just as valid.

Question 2

The Flavr Savr Tomato was first sold in 1994. This was a genetically modified tomato that contained an antisense gene. The antisense gene was a mirror image of the gene in the tomato that codes for polygalacturonase (PG), an enzyme that causes softening of the tomato as it ripens. The antisense gene transcribed mRNA that could base-pair with the mRNA from the PG gene.

(a) (i) Figure 2 shows a section of the mRNA transcribed from the PG gene. Annotate the figure to show the bases that would be present in the complementary strand of mRNA from the antisense gene. (AO1)

(1 mark)

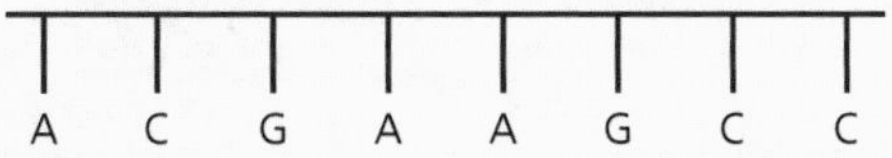

A   C   G   A   A   G   C   C

Figure 2

(ii) This antisense gene reduced the production of the PG enzyme to a very low level. Explain why. (AO2)　　　　　　　　　　　　(2 marks)

(iii) Suggest an advantage of reducing PG activity during tomato ripening. (AO3)　(2 marks)

Student A

(a) (i) UGCUUCGG

(ii) When the antisense mRNA binds to the PG mRNA it is double-stranded, so it can't bind to the ribosome and there are no exposed bases for tRNA to bind to.

(iii) If the tomato doesn't go soft as quickly, it is less likely to be damaged when being transported and so fewer tomatoes will be thrown away. This makes the tomatoes more profitable.

🅮 (a) (i) Full marks here — this is correct.

(ii) Again this is full marks. The question says that the two kinds of mRNA bind, but the student takes this further by explaining that it will not be able to attach to the ribosome and tRNA will not be able to bind.

(iii) This is also worth full marks as it is linked to damage to tomatoes and therefore profitability.

Student B

(a) (i) TGCTTCGG

(ii) The mRNA that codes for PG will bind to the mRNA from the antisense gene, so the mRNA cannot be translated.

(iii) If the tomatoes don't go soft quickly they are less likely to be damaged, so fewer will get wasted.

🅮 (a) (i) No marks here as the student has forgotten this is RNA, so A pairs with U not T.

(ii) The student has told us that the two kinds of mRNA bind together, but there are no marks for this as this is in the question. The student has not explained why the mRNA cannot be translated, so there are no marks here.

(iii) The student gets 1 mark here for understanding that fewer tomatoes will be wasted, but this is not linked to profitability, so the second mark cannot be awarded.

(b) When the FDA approved the Flavr Savr tomato, they stated that special labelling for the modified tomatoes was not necessary because they have 'the essential characteristics of non-modified tomatoes'. Evaluate this statement. (AO3)　　　　　　　　　　　　　　　　　　(4 marks)

Student A

(b) This is true in one way because the Flavr Savr tomato still has all the genes that a non-modified tomato has ✓. However, it does have one extra gene added, the antisense gene ✓. This just blocks one gene — the gene for PG — so it should not affect anything else in the tomato such as nutrition ✓. However, it would seem sensible to label the product so that people can choose for themselves whether to buy it or not ✓.

ⓔ The student has made four valid points, so gets full marks. Also the student does have points 'for' and 'against', so this is worthy of full marks.

Student B

(b) A new gene has been added so the tomatoes should be labelled as people have the right to know ✓. It is possible that the PG gene affects something else in the tomato as well as softening, and by adding this gene another gene is inactivated or 'switched on' ✓. It is unnatural to do this to a tomato.

ⓔ This student gets 2 marks. Two valid points are made. The comment 'It is unnatural' not only gets no marks, but is not a valid argument unless developed and a reason for the opinion given. This student has only given points against the argument. More marks are available if you address both sides of the argument.

Question 3

Some students surveyed the plants in a meadow. They sampled the meadow using 0.25 m² quadrats. The table shows their results.

Species	% cover							
Yorkshire fog grass	15	10	5	20	10	30	20	10
Timothy grass	70	60	50	60	60	50	40	70
Plantain	0	5	10	5	0	0	10	10
Buttercup	5	0	10	0	5	0	10	0
Dock	0	0	5	0	0	5	0	0
Clover	5	5	5	0	10	0	10	0
Dandelion	5	10	0	10	10	10	5	10
Bare ground	0	10	15	5	5	5	5	0

(a) (i) Suggest a suitable method that the students should have used to select the positions for the quadrats. (AO1)   (2 marks)

   (ii) The students did not use a transect line. Suggest why. (AO3)   (2 marks)

Student A

(a) (i) They should use a random number generator ✓ to give them random *x,y* coordinates then place the quadrat at that point ✓.

(ii) The meadow would have been fairly uniform ✓, so it should be surveyed using random quadrats. A transect line is only useful when there is a gradient or a change in vegetation ✓.

ⓔ This student gets full marks, for not only saying that the quadrats should be placed at random but for giving a clear method. Similarly, the student gives a thorough explanation of when a transect is appropriate and when random sampling is appropriate.

Student B

(a) (i) The meadow could be divided up into a grid ✓ and squares to sample could be chosen using random numbers ✓, or they could sample every fifth square.

(ii) Transects are used to sample along a line.

ⓔ This student gets both marks in (i) and, indeed, mentions that systematic sampling is a valid alternative to random sampling. This would have gained a mark if the answer had not got both marks already.

In (ii) there are no marks because the student has not answered the question. The student needs to explain why a transect is not appropriate in this case.

(b) Describe how to measure percentage cover using a quadrat. (AO1)   (2 marks)

Student A

(b) The student estimates how much of the quadrat is occupied by each species using the small squares ✓. The total should add up to 100% ✓.

ⓔ Both marks for a fully correct answer.

Student B

(b) The student uses the small squares to see roughly how many squares are filled with each species ✓.

ⓔ 1 mark here.

(c) (i) Could the students use these data to calculate a species diversity index? Explain your answer. (AO3)   (1 mark)

(ii) The students used these data to say that the species richness of the meadow was 5. Is this a valid conclusion? Explain your answer. (AO3)   (2 marks)

Student A

(c) (i) They couldn't because they haven't counted the number of each species, just percentage cover.

(ii) This is the species richness of the quadrats they sampled ✓, but they didn't sample many quadrats and there might be another species present in the field if they took a larger sample ✓.

ⓔ Full marks for both parts of this question.

Student B

(c) (i) Yes

(ii) Yes because species richness is the number of species present ✓.

ⓔ No marks in (i) — in general, you do not get a mark for yes or no, without qualifying the answer. The student should understand what is needed to calculate a diversity index and then think about whether the required data are available. In (ii) the student gets 1 mark for knowing what species richness is. For a second mark the student should say that this assumes the 8 quadrats are representative of the whole meadow, or say that there might be more species present if more quadrats had been sampled.

Question 4

A student extracted some DNA from some of her cheek cells. She mixed the cells with some detergent and a protease enzyme. Then she added cold ethanol to precipitate out the DNA.

(a) Suggest why she used:

    (i) detergent (AO2)   (1 mark)

    (ii) a protease enzyme (AO2)   (1 mark)

Student A

(a) (i) To break down the cell membrane ✓ and nuclear membrane as they are made mainly of phospholipids.

(ii) To digest the histone proteins that surround the DNA ✓.

ⓔ Full marks here for two fully correct answers.

Student B

(a) (i) To dissolve fats

(ii) To digest proteins

ⓔ No marks here. The detergent does disrupt lipids, but this must be linked to the investigation and the need to break down the membrane. Similarly in (ii), the student needs to say which proteins need to be digested.

The student carried out an analysis on the DNA she had extracted. She added a restriction enzyme to the DNA in a buffer solution and incubated the mixture at 37°C for 20 minutes.

(b) Explain why the mixture:

   (i) was incubated at 37°C for 20 minutes (AO1)   (2 marks)

   (ii) included a buffer solution. (AO1)   (1 mark)

Student A

(b) (i) This is the optimum temperature for the restriction enzyme ✓ and this gave it enough time to cut the DNA at all the restriction sites it could ✓

   (ii) This provides the optimum pH for the restriction enzyme ✓

ⓔ Full marks here for two detailed answers.

Student B

(b) (i) This gave the enzyme time to digest the DNA ✓.

   (ii) This keeps the pH stable.

ⓔ Only 1 mark in (i) because there is no reference to optimum temperature. In (ii), although this is the function of a buffer, the use of buffer is not related to the investigation and enzyme, so there is no mark.

After this, the student pipetted a sample from this mixture into well B at one end of an agarose gel in an electrophoresis tank. In well A she pipetted a sample of DNA containing a mixture of fragments of known length. She carried out gel electrophoresis. Figure 3 shows the appearance of the gel after electrophoresis had been carried out and the DNA fragments stained.

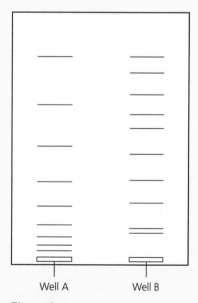

Well A   Well B

Figure 3

(c) (i) Applying an electric current across the gel causes the DNA fragments to separate. Explain how. (AO1)   (2 marks)

(ii) Suggest the purpose of the DNA fragments in well A. (AO3)   (2 marks)

(iii) Explain why a stain was needed. (AO3)   (1 mark)

Student A

(c) (i) DNA is negatively charged ✓, so it moves towards the positive electrode ✓.

(ii) These are a known length, so we can compare how far they travel with the fragments in the cheek cell DNA ✓. This helps us to work out how long the cheek cell DNA fragments are ✓.

(iii) So that you can see where the DNA bands are because DNA is colourless ✓.

🅔 Full marks for a fully correct set of answers.

Student B

(c) (i) The smaller pieces of DNA move further along the gel.

(ii) So you can work out how long the pieces of DNA in the cheek cell DNA are ✓.

(iii) To show up the DNA which is invisible otherwise ✓.

🅔 In (i) there are no marks because this is not linked to the electric current. The student needs to mention the charge on DNA and how it moves to the anode. In (ii) there is 1 mark but not a second as the student does not explain how the DNA helps to work out the length of the cheek cell fragments. In (iii) the student gains the mark for a correct answer.

Question 5

Tuberculosis (TB) is a bacterial disease that primarily infects the lungs. Cattle that are infected with bovine TB have to be destroyed. Many people believe that one source of infection for cattle is badgers.

In an investigation, a group of scientists tested a number of different animals for the bacterium *Mycobacterium bovis* that causes bovine TB. They carried out PCR on DNA from different samples, using primers that bind to *Mycobacterium bovis* DNA. The table shows some of their results.

Animal	Number tested	Number of samples giving positive test for M. bovis DNA
Yellow-necked mouse	268	0
Wood mouse	1338	0
Bank vole	1307	1
Rat	76	0
Common shrew	272	0
Rabbit	202	0
Grey squirrel	189	0
Badger	43	3
Dog	21	0
Cat	42	0

(a) (i) Calculate the percentage of badgers with a positive test for *Mycobacterium bovis*. Show your working. (AO2)

(2 marks)

Student A

(a) (i) 3/43 × 100% = 6.98%

e Full 2 marks for a correct answer.

Student B

(a) (i) 6.976%

e This also gets both marks, but note that this student did not show any working. If the student had made an arithmetic mistake, they would have lost the opportunity to get a mark for the correct method.

(ii) The scientists tested mucus from the trachea, urine and faeces. Suggest why these samples were chosen. (AO3)

(2 marks)

Student A

(a) (ii) It is a lung disease, so the bacteria should be detectable in the trachea if present ✓. Urine and faeces get left in places where the animal has been, such as on grass that cattle eat, so if TB passes from wild animals to cattle this is a source that needs testing ✓.

e This gets 2 marks for two valid reasons.

Student B

(a) (ii) These samples can be tested without causing any harm to the animals that are tested ✓.

ⓔ This gains 1 mark for a sensible idea, but the student has not thought about the aim of the investigation — the scientists also need valid results so reasons why these are valid samples to test should be given.

(b) Many other sources of DNA would have been present in the samples obtained. Explain how the use of primers ensured that only *Mycobacterium bovis* DNA was isolated. (AO2)　　　　　　　　　　　　　　　(2 marks)

Student A

(b) The primers used would be complementary to the start and end of a sequence of DNA that is found in *Mycobacterium bovis* ✓. They will not bind to any other DNA, so other kinds of DNA will not be amplified using PCR ✓.

ⓔ This gets both marks for a fully correct answer.

Student B

(b) The primers will only bind to the DNA of the TB bacteria because of complementary base-pairing ✓.

ⓔ This gets 1 mark for the specificity of the primers, but there is no second mark as it is not linked to this DNA being amplified by PCR.

(c) A journalist said that these results support the idea that badgers spread TB to cattle and should be culled. Evaluate this statement. (AO3)　　　　　(4 marks)

Student A

(c) The results do show that badgers can carry the TB bacteria but that almost all the other animals tested did not carry TB ✓. However, most of the badgers tested did not have the TB bacteria ✓. This investigation does not give any data for numbers of cattle in the area with TB ✓ and even if these badgers do have TB we don't know that cattle have caught TB from the badgers and not another way ✓.

ⓔ Full marks here for four arguments, addressing both sides of the argument.

Student B

(c) This shows that, apart from one bank vole, the only animals that have TB are badgers so this supports the journalist ✓. Also a very large sample of animals was tested, so these results should be reliable ✓. However, this study doesn't tell us that culling badgers is the best way to deal with the problem compared with vaccinating badgers or cattle ✓.

ⓔ This gets 3 marks for three valid points. Again this student has given arguments on both sides which is good.

(d) (i) The scientists used the capture-mark-recapture method to estimate the population of each species of animal in the area they were testing. Explain why it was important to estimate the total population of each species. (AO3)

(2 marks)

Student A

(d) (i) The scientists needed to be sure they had tested a reliable sample of animals in the area ✓. This means they have to be sure they have tested a high proportion of the population ✓.

ⓔ This gets both marks for a fully correct answer.

Student B

(d) (i) The scientists need to test a representative sample of the population ✓ for reliable and valid results ✓.

ⓔ This answer is also fully correct, so worth 2 marks.

(ii) In one area they captured 37 common shrews. They marked the animals by clipping their fur in one area and released them. The next evening they captured 43 common shrews, of which 12 were marked. Calculate the population of common shrews in that area. Show your working. (AO2)

(3 marks)

Student A

(d) (ii) $\dfrac{\text{number marked and released}}{\text{total population}} = \dfrac{\text{number recaptured}}{\text{size of second sample}}$

$\dfrac{37}{\text{population}} = \dfrac{12}{42}$

$\text{population} = \dfrac{37 \times 42}{12}$

$= 129.5 = 130$ to nearest whole number

ⓔ This gets 3 marks for a fully correct answer.

Student B

(d) (ii) $\dfrac{\text{number marked and released}}{\text{total population}} = \dfrac{\text{number recaptured}}{\text{size of second sample}}$

$\text{population} = \dfrac{37 \times 12}{42}$

$= 10.57$

ⓔ The student gets 1 mark for the first line of working, but he/she has not rearranged the equation properly, so gets no further marks. The student should have realised this is an incorrect answer because the scientists caught 37 and 42 animals, so the population size must be a lot larger than 11!

Question 6

Write an essay on one of the following:

(a) The roles of inorganic ions in living organisms.

(b) The biological importance of proteins.

(25 marks)

Student A

(a) Phosphates are important in the molecule ATP, adenosine triphosphate. ATP is produced in respiration and in the light-dependent stage of respiration by adding a phosphate ion to adenosine diphosphate. This is catalysed by the enzyme DNA synthase. When a phosphate breaks off adenosine triphosphate, energy is released that can be used in many processes, such as spindle formation in mitosis, muscle contraction, protein synthesis or active transport. Phosphate is also important in nucleotides. Every nucleotide in DNA or RNA contains a phosphate group, which joins to the pentose sugar in the adjacent nucleotide to form a phosphodiester bond, forming the backbone of the polynucleotide. The phosphate group gives the nucleotides a negative charge, which means that fragments of DNA move towards the positive electrode in gel electrophoresis. Phosphate groups are added to glucose at the start of glycolysis, from ATP, to phosphorylate it and make it more reactive.

Hydrogen ions are important in biology as they change the pH in living organisms. The higher the concentration of hydrogen ions, the lower the pH. When hydrogen ions build up in the blood, which happens when the cells are respiring actively and producing more carbon dioxide, the hydrogen ions attach to the haemoglobin in red blood cells. This changes the affinity of the haemoglobin for oxygen, and causes the haemoglobin to release more oxygen. This is called the Bohr shift and it enables aerobic respiration to continue for longer. When hydrogen ions build up in a cell, this can affect the hydrogen and ionic bonds holding protein tertiary structures in shape. If too many of these bonds break, the proteins can become denatured. Enzymes are proteins so this can change the shape of the enzymes' active sites and stop their activity. Hydrogen ions are also important in electron transfer chains in the mitochondrion. Reduced coenzymes bring hydrogen atoms to protein carriers in the inner mitochondrial membrane. These split into hydrogen ions and electrons. The hydrogen ions build up in the space between the inner and outer mitochondrial membranes. When they pass back through the stalked particles into the mitochondrial matrix, they release enough energy for ATP synthase to make ATP.

Iron ions are found in haemoglobin in the haem groups and it is the haem group that binds the oxygen. Chlorophyll in plant chloroplasts has a similar structure to haem, but this has a magnesium ion at its centre. Inorganic metal ions can act as enzyme cofactors which means they are needed, as well as the enzyme, for an enzyme-catalysed reaction to occur.

Sodium ions are important in action potentials, as well as potassium ions. When a neurone is at rest, the sodium–potassium pump actively transports three sodium ions out of the cell in return for two potassium ions coming in. This creates a potential difference, so that the inside of the neurone is negatively charged compared to the outside. When a stimulus is detected by a receptor, this leads to the neurone membrane being depolarised. Sodium-gated channels open and sodium ions enter the neurone down their electrochemical gradient, causing the potential difference across the membrane to be reversed. Then the sodium channels close and the potassium-gated channels open. Potassium ions leave the membrane down their electrochemical gradient, repolarising the membrane. The sodium–potassium pump redistributes the ions. At the synapse, calcium ions enter the presynaptic neurone causing vesicles of neurotransmitter to fuse with the presynaptic membrane. The neurotransmitter diffuses across the cleft and fits into receptor proteins on the postsynaptic membrane. This opens sodium-gated channels. If enough sodium ions enter the postsynaptic neurone to overcome the threshold value, an action potential is set up in the postsynaptic neurone.

Sodium ions are also important in the co-transport of glucose and amino acids into gut epithelial cells. Sodium ions are actively transported out of the gut epithelial cells into the blood capillaries, creating a low sodium ion concentration inside the cells. Glucose in the gut lumen fits into a specific receptor protein in the membrane of the gut epithelium cell, along with a sodium ion. They are co-transported into the cell. The glucose enters the blood capillary by facilitated diffusion. Amino acids enter the cell in a similar way except that they are co-transported by sodium ions using a different receptor protein in the cell membrane.

Calcium ions are also important in muscle contraction when they inhibit troponin and allow muscle contraction.

ⓔ This is a good essay. There are no significant errors, and the student uses correct scientific terminology. A range of different ions are used as examples, and the role of the ions is clear as well as just saying where the ions are used. The essay contains good A-level detail from different parts of the specification. This therefore gets 3 marks for use of language, 3 marks for breadth and 3 marks for relevance as there is no irrelevance here. For scientific content this gets 12 marks as it does not have very much beyond the A-level specification, although the references to magnesium in chlorophyll and metal ions in cofactors are not directly on the specification. In total, therefore, this gets 21 marks out of 25.

(b) Proteins are important in biology in many ways and I am going to write about some of these. I am going to mention how proteins are made and then how they are important in our diet, as enzymes and in membranes.

Proteins are polymers made when two amino acids join by a condensation reaction. An amino acid has an amino group, a carboxylic acid group and a side chain called an R group. The R groups are different in different amino acids. When they join to form a dipeptide, a hydrogen from the amino group of one amino acid joins with the –OH from the carboxylic acid group of the other amino acid, and forms water. There is a bond between the N of one amino acid and the C of the next and this is a peptide bond. This carries on until there is a polypeptide chain. The polypeptide chain folds up, with hydrogen bonds forming between different R groups, to form a secondary structure. The secondary structure folds up to form a tertiary structure containing disulfide bridges, ionic bonds and hydrogen bonds.

Enzymes are proteins. Proteins have a complex tertiary structure that means they have an active site that is specific to a substrate. They work by induced fit. When the substrate binds to the active site, the active site changes shape, so that the substrate fits it exactly. This lowers the activation energy needed for the reaction to occur. The product is released from the active site and the enzyme can be used over and over again. If the pH changes too much away from the optimum, or the temperature rises too high, the enzyme will be denatured because the ionic and hydrogen bonds holding the protein in its tertiary structure will break. Enzymes are highly specific because only the right substrate will fit into the active site.

Haemoglobin is a protein that transports oxygen in red blood cells. Oxygen binds to the haem group. When one oxygen binds, the haemoglobin changes shape so that the next oxygen binds more easily. There are different kinds of haemoglobin in different organisms. Fetal haemoglobin has a higher affinity for oxygen than the mother's haemoglobin so it can become fully saturated with oxygen at the placenta. Diving animals like seals have a haemoglobin with a very high affinity for oxygen so they can survive underwater without breathing for about 30 minutes. The Bohr shift is when carbon dioxide causes the haemoglobin to dissociate more readily. This is useful when an individual is respiring a lot as it means the cells can respire aerobically for longer.

There are proteins in the cell membrane that act as carriers and channels. These are specific to one or maybe a couple of different molecules because of their tertiary structure. In facilitated diffusion, the molecule binds to a carrier and the carrier changes shape, releasing the molecule the other side of the membrane. This happens down a concentration gradient. In active transport a similar change happens, except that ATP is needed for the carrier to change shape and the molecule is transported against its concentration gradient.

DNA is a protein. It is made of monomers called nucleotides that form a double helix structure. Each nucleotide has a base in it and the four bases are adenine, guanine, thymine and cytosine. A pairs with T and C with G. This means that when DNA replicates semi-conservatively the new strand formed is complementary to the old strand. Three bases in DNA is a triplet that codes for one amino acid. DNA is present in the nucleus of cells and forms our genes. These genes code for proteins. In a recent New Scientist article it reported that DNA from a retrovirus was present in Neanderthals and in earlier humans as well as modern humans. This gives information about how humans evolved and shows that the virus infected humans before modern humans evolved.

Other examples of proteins are antibodies. Proteins are important in action potentials where there are sodium and potassium channels. Muscles are made of the proteins actin and myosin. There are many more important roles of proteins and I have discussed some of them but there are many more.

ⓔ This essay has some good biology in it, but does not get a top mark. The first paragraph is an introduction. There is no need for an introduction in an essay like this — it does not have any detailed A-level information so it gets no marks. This student has wasted time writing an introduction that gets no marks. The second paragraph contains some excellent biology, but it is not relevant. The essay title is 'the importance of proteins' not 'write everything you know about proteins'. The structure of proteins is not relevant. Remember, if you were asked about the importance of computers in everyday life, you might talk about emails, the internet or using spreadsheets, but you would not give details of how to make a computer. Paragraphs 3, 4 and 5 are all relevant and they do contain some detailed and relevant A-level biology. However, paragraph 5 contains a fundamental error — the student believes that DNA is a protein — and the whole paragraph is irrelevant. The student does attempt to put in some information that is beyond A-level, but as this is not relevant, it does not count. Finally, the student writes a conclusion. This is also a waste of time as it has no detailed A-level biology in it. It would have been better if the student had spent the time writing a good paragraph about antibodies or proteins in muscle instead of writing the introduction and conclusion. So, there are three good relevant paragraphs with no significant errors. This gives a content mark of 8, with 2 marks for breadth, 1 mark for relevance and 3 marks for quality of language, making 14 marks out of 25.

Required practical 12 answers

1 Using a random number table to provide coordinates.

2 There is no correlation between height of buttercup and soil pH.

3 Spearman's rank (or any other) correlation coefficient. For Spearman's rank, you have to:
   - Rank each set of data.
   - Find the difference between the two ranks for each pair of values (you can ignore whether the difference is positive or negative).

- Square the difference in ranks.
- Sum the squares of the differences.

Sample	Height of plant/cm	Rank 1	pH of soil	Rank 2	Rank difference (rank 2 – rank 1 = d)	Rank difference squared (d^2)
1	27.5	2	6.3	9	7	49
2	18.0	7	7.2	3	24	16
3	11.0	10	7.6	1	29	81
4	22.5	3	6.4	8	5	25
5	29.5	1	6.8	6.5	5.5	30.25
6	16.5	8	7.0	5	23	9
7	19.0	5.5	6.0	10	4.5	20.25
8	13.5	9	6.8	6.5	–2.5	6.25
9	21.0	4	7.4	2	22	4
10	19.0	5.5	7.1	4	–1.5	2.25
						$\sum d^2 = 243$

The formula for Spearman's rank correlation is

$$r_s = 1 - \frac{6\sum d^2}{n(n^2 - 1)}$$

where r_s = Spearman's rank correlation coefficient

n = number of pairs of items in the sample

d = the difference between each pair of ranked measurements

$$r_s = 1 - \frac{6 \times 243}{10^3 - 10}$$

$$= 1 - \frac{1458}{990} = 1 - 1.47 = -0.47$$

4 The calculated value is less than the critical value, so the null hypothesis must be accepted.

5 Any two sensible ideas are valid here, e.g. water availability, light intensity, soil compaction, soil nutrients. It is very difficult to control these, but the student could have improved the investigation by measuring these variables at each sampling point. If they are similar, they are unlikely to affect plant height, but if they are variable then this factor might be affecting the height of the buttercups and not just pH.

Knowledge check answers

Knowledge check answers

1 1B, 2E, 3H, 4A, 5C, 6G, 7D, 8F

2 The parents are normal so they must both have the normal dominant allele (N). However, their baby has cystic fibrosis, which is recessive, so the baby must have received a recessive allele from each parent. Therefore we know the parents' genotypes are Nn.

Parent phenotypes   normal   normal
Parent genotypes   Nn   Nn
Gametes   N n × N n
Offspring genotypes

	N	n
N	NN	Nn
n	Nn	nn

Offspring phenotypes
NN : Nn : nn
Normal : normal, carrier : cystic fibrosis
1 : 2 : 1
Therefore there is a 1 in 4, ¼ or 25% chance that their next baby will also have cystic fibrosis.

3 Parent phenotypes   roan   roan
Parent genotypes C^RC^W C^RC^W
Gametes C^R C^W × C^R C^W
Offspring genotypes

	C^R	C^W
C^R	C^RC^R	C^RC^W
C^W	C^RC^W	C^WC^W

Offspring phenotypes   roan : red : white
2 : 1 : 1

4 Yes, if she inherits the allele from both parents.
Parent phenotypes
   normal woman, carrier   haemophiliac male
Parent genotypes X^HX^h X^hY
Gametes X^H X^h × X^h Y
Offspring genotypes

	X^h	Y
X^H	X^HX^h	X^HY
X^h	X^hX^h	X^hY

X^hX^h is a haemophiliac girl.

5 Parent phenotypes
   colourblind female   normal male
Parent genotypes X^bX^b X^BY
Gametes X^b X^b × X^B Y
Offspring genotypes

	X^B	Y
X^b	X^BX^b	X^bY
X^b	X^BX^b	X^bY

Offspring phenotypes
   all daughters normal vision (but carriers) and all sons red–green colourblind
So the chance they will have a child with red–green colourblindness is 50%.

6 Parent phenotypes
   tortoiseshell female   ginger male
Parent genotypes X^GX^B X^GY
Gametes X^G X^B × X^G Y
Offspring genotypes

	X^G	Y
X^G	X^GX^G	X^GY
X^B	X^GX^B	X^BY

Offspring phenotypes
   1 ginger female : 1 tortoiseshell female: 1 ginger male: 1 black male

7 a

Genotype	Phenotype
Cc	Dark grey
$c^{ch}c^h$	Chinchilla
c^hc	Himalayan
cc	White

b Parent phenotypes   dark grey   chinchilla
Parent genotypes   Cc $c^{ch}c$
Gametes   C c × c^{ch} c
Offspring genotypes

	c^{ch}	c
C	Cc^{ch}	Cc
c	$c^{ch}c$	cc

Offspring phenotypes
   2 dark grey : 1 chinchilla : 1 white

8 Parent phenotypes
   red–green colourblind non-rolling man
   non-colourblind tongue-rolling woman
Parent genotypes X^bYrr X^BX^BRr
Gametes X^br Yr × X^BR X^Br
Offspring genotypes

	X^br	Yr
X^BR	X^BX^bRr	X^BYRr
X^Br	X^BX^brr	X^BYrr

Offspring phenotypes
   None of the offspring are colourblind, but half the males and half the females are tongue-rolling while the other half are non-rollers.

9 Parent phenotypes   green   green
Parent genotypes   AaBb   AaBb
Gametes AB Ab aB ab × AB Ab aB ab

Offspring genotypes

	AB	Ab	aB	ab
AB	AABB	AABb	AaBB	AaBb
Ab	AABb	AAbb	AaBb	Aabb
aB	AaBB	AaBb	aaBB	aaBb
ab	AaBb	Aabb	aaBb	aabb

Offspring phenotypes
9 green: 3 blue: 3 yellow: 1 white

10 64% have unattached earlobes, so we know that
$(p^2 + 2pq) = 0.64$
Therefore $q^2 = 0.36$
So $q = \sqrt{0.36} = 0.6$
$p + q = 1$ therefore $q = 1 - 0.6 = 0.4$
$2pq = 2 \times 0.6 \times 0.4 = 0.48$
so 48% of the Chinese population are heterozygous for earlobe attachment.

11 Crossing-over during prophase I produces new combinations of alleles. Also, random assortment/independent segregation of chromosomes in the first division of meiosis, and of chromatids in the second division of meiosis, results in new combinations of alleles.

12 **a** When antibiotics are used, most bacteria are susceptible and die. Those bacteria that, by chance, have a mutation that gives them resistance to the antibiotic are more likely to survive. Therefore they will reproduce and pass on their allele to their offspring. Therefore the allele for antibiotic resistance increases in frequency in the population.

   b Babies born at a low birthweight are less likely to survive as they are not strong enough. However, large babies are also less likely to survive as they may face difficulties during birth. Therefore the babies that are most likely to survive to adulthood and pass on their alleles are those of intermediate birthweight.

13 The original finch population showed variation in beak size and shape. In the area where cacti grew, any finch with a beak that was better adapted to eat cacti survived longer, because it got more food. Therefore it reproduced more and passed on its favourable alleles. A mutation may have occurred that was more advantageous and a finch with this advantageous allele would pass it on. Eventually the allele frequency for alleles conferring a short, stout and strong beak would increase in this population. In the area where trees were common, finches with a beak that enabled them to eat insects from tree bark would survive longer. This means that alleles that produced a long, thin beak would be passed on and increase in frequency in that population. Eventually, in one population a mutation might occur that led to reproductive isolation, e.g. a change in the courtship ritual. As a result, the two populations would be

unable to interbreed to produce fertile offspring, and they would now be separate species.

14 This is an example of the founder effect. Among the small number of original colonists there would have been one or two individuals who carried the allele for Huntington's disease, giving this allele a much greater frequency in this small population than it would have had in the original population from which the colonisers came. As these colonisers interbred, the allele was passed on and remains in a higher frequency than in most other populations.

15 The graph showing the two species grown separately indicates that both species can not only survive in these culture conditions, but can reproduce. However, when grown together *P. aurelia* increases in population size while *P. caudatum* decreases to zero. This shows that there is interspecific competition for a resource, e.g. food supply, and *P. aurelia* is better adapted to the conditions than *P. caudatum*.

16 This method is neither random nor systematic. No matter how hard you try, you will have some idea of which direction the quadrat will land in, so there is bias introduced.

17 total population = number of animals marked and released × total number in second sample/number marked animals in second sample
= 23 × 28/9
= 71.56
As the answer must be a whole number, the estimated carp population is 72.

18 **a** There is no difference in the number of worms between fields A and B.

   b $\chi^2 = 56.38$
   There is 1 degree of freedom. The critical value for $p < 0.05$ and 1 degree of freedom is 3.84 (using Table 4 on p. 14). The calculated value is much higher than this, so we can reject the null hypothesis. There is a significant difference between the number of earthworms in each field. There is a less than 0.01 probability of obtaining these results by chance.

Number of worms	Observed (O)	Expected (E)	(O – E)	(O – E)^2	(O – E)^2/E
Field A	143	125	18	324	2.59
Field B	207	125	82	6724	53.79
					56.38

19 During interphase (in the synthesis phase)

20 The DNA code is degenerate, which means there is more than one codon for many amino acids. If the substitution affects the third base in the triplet, this may change it from one triplet to a different triplet for the same amino acid.

21 The new amino acid will have a different R-group, which means that the hydrogen, ionic and disulfide bonds that hold the polypeptide in its secondary and tertiary structure may form in a different place. This means the protein will be a different shape and may not function correctly.

22 Multipotent because they produce blood cells but no other kind of cell. They are not unipotent because they can produce different kinds of blood cell.

23 In the direction of transcription, they are located in front of the gene they regulate.

24 They have a shape that is complementary to part of the RNA polymerase molecule.

25 Only certain cells contain the protein receptor that oestrogen binds to.

26 a Yes (mRNA is being made), **b** no — mRNA is destroyed before it can reach the ribosome.

27 If cancer is detected early, it is possible to remove the whole primary tumour before it has had a chance to spread to other parts of the body.

28 Genes that are under a great deal of selection pressure are likely to show the most variation between organisms of the same species. For example, an antigen that stimulates an immune response will lead to the host developing specific antibodies. If another parasite, as the result of a mutation, has a slightly different shape of antigen, it is more likely to survive the immune response of the host, and pass on its allele to its offspring.

29 The main advantage is that these methods produce DNA without introns, so the gene can be transcribed in any organism including a prokaryote. (There are other possible benefits too. For example, it is difficult to locate the gene you require in a donor cell without the use of gene probes.)

30 The primer is needed for DNA polymerase to attach. Two different primers are needed as the base sequence of DNA is different at the ends of each strand.

31 This is so the enzyme can withstand the heating to 95°C. If it was not thermostable, it would denature, and new enzyme would need to be added every cycle.

32 64 (number doubles every cycle — so 2,4,8,16,32,64).

33 All the cells in a colony have come from a single bacterial cell dividing asexually, so they must be genetically identical.

34 If excess probe was not washed off, it would remain on the paper and cause fogging. This would give the misleading impression that a colony of bacteria contains the gene of interest when it does not.

35 'In vitro' means 'in glass'. Although PCR involves plastic tubes, the term 'in vitro' has come to mean any technique carried out using laboratory apparatus and not living cells. 'In vivo' means 'in life' and so copying DNA using a bacterial cell comes in this category.

36 Even if the gene gets into the cells, it only changes that cell. When the cell dies, it is replaced by a cell with the 'faulty' gene only. Therefore the treatment has to be repeated to get the 'healthy' gene into these new cells.

37 DNA contains phosphate groups which are negatively charged, so it moves towards the positive electrode.

38 In PCR, a probe is used that binds to human DNA but not bacterial DNA. Therefore it is only the human DNA that is amplified.

Index